KT-414-517

SUPERGUIDES

MUSHROOMS AND TOADSTOOLS

DEREK REID

KINGFISHER BOOKS

Kingfisher Books, Grisewood & Dempsey Ltd,
Elsley House, 24–30 Great Titchfield Street,
London W1P 7AD

This edition published in 1989 by Kingfisher Books.
Material in this book was first published in 1980
in the Kingfisher Guide series

British Library Cataloguing in Publication Data
Reid, Derek, *1927–*
Mushrooms and Toadstools.
1. North-western Europe. Fungi
I. Title II. Series
589.2'094

ISBN 0-86272-485-6

All illustrations by Bernard Robinson except p.4, Sean Milne

Edited by Stuart Cooper
Designed by Millions Design
Printed in Hong Kong

CONTENTS

Introduction 4
Agarics 6
Boletes 35
How to Take a Spore Print 38
Glossary 38
Index 39

INTRODUCTION

The term 'mushroom' was, until fairly recently, restricted to the field mushroom and the horse mushroom (*Agaricus campestris* and *Agaricus arvensis* respectively) among the wild species, but included the cultivated mushroom (*Agaricus bisporus*). All other gill-bearing fungi, sometimes referred to as the agarics, were known as toadstools. However, under the influence of American usage, the word mushroom has been broadened to cover not only all agarics but all the larger, fleshy fungi.

Fungi have a very varied appearance, ranging from patches on wood, through brackets, coral-like tufts, simple clubs and rosettes, to cauliflower-like structures or centrally or laterally stalked fruitbodies. The texture varies from woody, leathery or fleshy to gelatinous and the surface may be either smooth, velvety, hairy or scaly and either dry or glutinous. The fertile underside may be smooth or have spines, pores or gills.

Fungi have certain features which clearly separate them from all other plants. Apart from differing in shape, they all lack the green pigment chlorophyll and are unable to manufacture their own carbohydrates by photosynthesis. Another distinction involves structure: fungi are all formed of branching threads or hyphae, which are quite different from the cells of other plants. Again, reproduction is different: fungi spread by minute spores, visible only under a microscope.

THE FUNCTION OF A MUSHROOM

A mushroom is only the reproductive part (known as the fruitbody) of a fungus and is produced on, and nourished by, an extensive network of hyphac. Fungi obtain their food by the breakdown of plant or animal material. They are, therefore, frequent in woodland rich in humus, but also grow in pastures, on heaths, sand-dunes, or mountaintops on wood, soil or dung. In such species, the network is a cobwebby mat of hyphae (known as a mycelium) which branches out amongst debris on the woodland floor, amongst dead or decaying grass, or in rotting wood. However, some fungi are parasitic, even of other fungi (for example *Boletus parasiticus*). In these species, the mycelium is found within the host tissue.

DEVELOPMENT OF THE FRUITBODY

Once a mycelium is established, it grows and eventually builds up sufficient food reserves for the fruitbody to form. For example, in the case of an *Amanita* species, dense knots of hyphae form at certain points and each of these eventually develops into a 'button-stage' fruitbody. If this were to be cut through vertically, the button would be seen to consist of a tiny cap with gills on the underside and a stalk. The gills are covered by a membrane stretching from near the apex of the stem to the margin of the young cap – the 'partial veil'. The entire developing fruitbody is enclosed within another membrane – the 'universal veil'. As the young fruitbody enlarges, the stalk elongates, pushing the cap upwards until it breaks the universal veil. If the surface of the cap is sticky or, if the membrane is fibrous, the cap slips through cleanly and the universal veil is left at the base of the stem as a sac-like sheath – the 'volva'.

The stem and cap continue to grow and, as the cap expands, the remnants of the universal veil become stretched and break up into

DEVELOPMENT OF THE FRUITBODY
(*Amanita phalloides* or 'Death Cap')

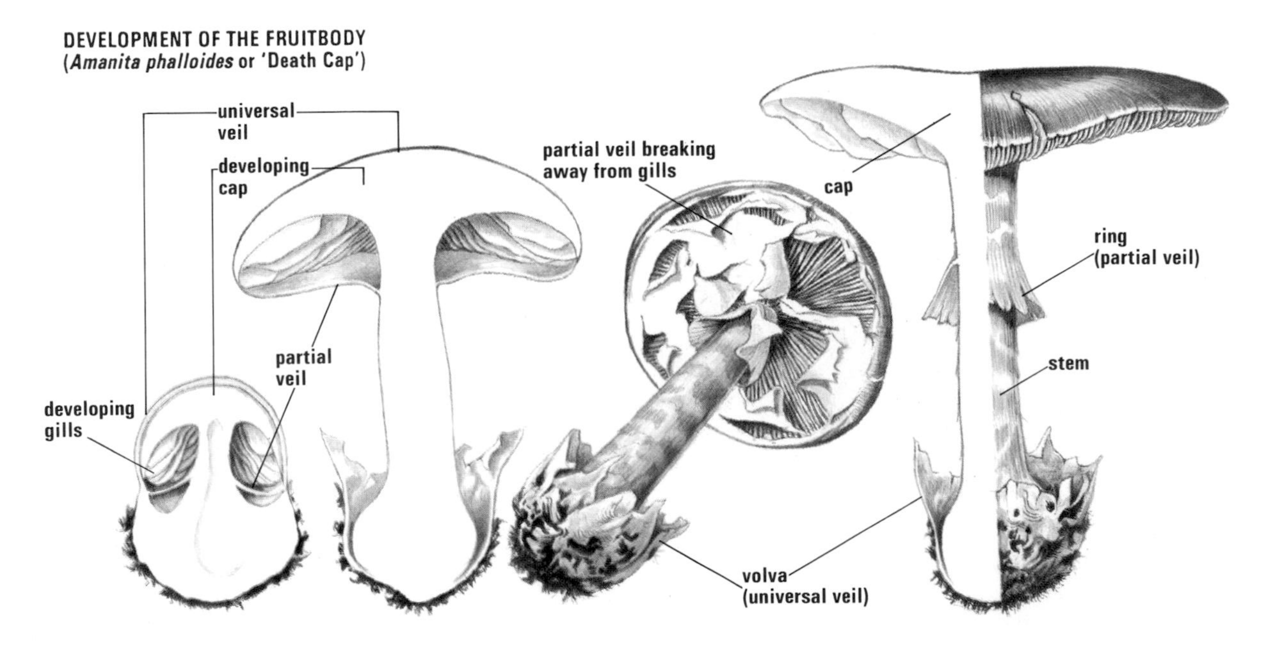

warts or mealy scales which gradually disperse towards the cap margin. The partial veil protecting the developing gills also becomes stretched and finally breaks away from the cap margin to fall back on the stem as a ring (annulus).

REPRODUCTION
The spores, which are the equivalent of seeds in flowering plants, are produced on the surface of the gills or within the pores on structures known as basidia. Spores are of constant size, shape and colour for every species and are often ornamented. They are very important in identification.

At maturity the spores are shot away for a short distance and then fall free of the gills or pores under their own weight, to be dispersed by air currents. On reaching a suitable habitat, the spores germinate, each producing a new mycelium, which must fuse with another mycelium of the same species in order to form fruitbodies. Fruitbody formation can then occur provided the mycelium has sufficient food material and that the temperature, humity and light requirements are met.

FUNGI IN COMMERCE
Fungi are of considerable economic importance. A number of species are parasitic on other plants and cause considerable losses to foresters, farmers and gardeners alike. For example, *Armillaria mellea* causes the death of many deciduous trees and conifers each year.

Balancing the harmful effects of parasitic fungi in forestry are other species, which form an association with the roots of trees. Food materials absorbed by the fungus from the soil became available to the tree while various substances manufactured by the tree become available to the fungus. The relationship is therefore mutually beneficial. It also has commercial applications, for when trying to establish plantations of trees in new areas or in a foreign country, it is often necessary to ensure the region is infected with the correct fungus.

FUNGI AS FOOD
The only safe way to learn how to distinguish edible from poisonous mushrooms is to go out with someone who can point out the edible species and the features by which they are recognised. In general, you should always avoid any white-spored species which have both a ring and a sac-like volva. If you think you have an edible field or horse mushroom (*Agaricus campestris* or *Agaricus arvensis*) you should ensure that the spore-print is purplish-black in colour and that there is a ring on the stem. Also, scratch the base of the stem and the cap and avoid any fruitbodies which bruise bright yellow.

Carry fungi you have collected in shallow open baskets, placing each collection in a paper bag or, if the species is very small or delicate, in tins lightly packed with moss to prevent damage. Avoid plastic bags as excess humidity causes rapid decay. Dig up specimens with a knife to ensure that any volval remains at the base of the stem are present. Finally, note the tree under or on which the fungus is growing, since this is an important feature in identification.

THE MUSHROOMS COVERED IN THIS GUIDE
Two groups of mushrooms are covered in this guide: the agarics and the boletes.

The agarics are fleshy fungi which bear gills. They grow on the ground in pasture or woodland, on dung, trees or on woody debris. The stem may or may not bear a ring or conspicuous zone of fibrils and may or may not show a sheathing, sac-like structure at its base.

Before attempting to identify an agaric it is essential to take a spore-print to check the colour of the spore powder. The procedure for doing this is described on page 38. Once the colour of the print has been established, you can then turn to the section of this book where the group of agarics with that colour of spore-print is dealt with.

When trying to assign an agaric to its correct genus, it is necessary to note the type of gill attachment to the stem, and for which there are several terms. *Free* means that the gills are not attached to the stem; *decurrent* is when the gills run down the stem; *adnate* is when the gill is attached by its entire width; *adnexed* when attachment is by less than its total width; *sinuate* when there is a notch in the gill where it joins the stem.

Mushrooms of the bolete group resemble the agarics in that they also have fleshy fruitbodies, with a cap and central stalk which may or may not bear a ring. The distinction can be seen on the underside of the cap. Whereas the agarics have gills, the boletes have a sponge-like structure with tiny pores. These are the openings of densely crowded tubes lined with basidia which produce the spores. With the exception of *Boletus parasiticus*, mushrooms of this group all grow on the ground in association with trees.

Inevitably, readers will find numerous species which are not described in this book, but most common species have been included. The author has tried to select only those species which can be identified without using a microscope.

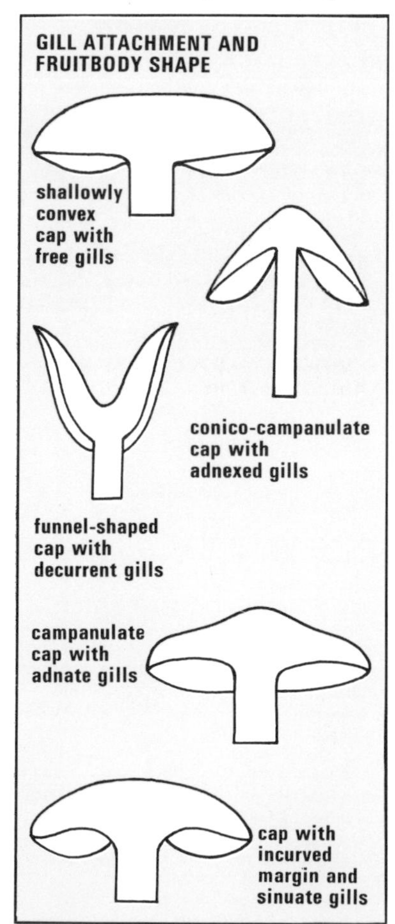

AGARICS

AMANITA

A genus of approximately 25 species, having in common a white spore-print, a volva at the base of the stem and usually, but not always, a membranous ring. The volva may be conspicuous and sac-like or reduced to a narrow rim or to concentric rings of scales. The cap is often ornamented with conspicuous warts, flat patches of tissue or hoary scales but when present these are superficial and are easily removed. The genus *Amanita* includes species which are amongst the most poisonous known agarics.

AMANITA PHALLOIDES
Death Cap

Cap: 6-9 cm diam., convex then shallowly convex; varying in colour from olive-green at the centre to yellowish-green nearer the smooth margin. The cap surface appears indistinctly radially streaky; usually naked, without trace of warts or scales.
Stem: 7-9 cm high; 1-1·5 cm wide; white, cylindrical or narrowed above and sheathed below in a conspicuous free-standing, white, sac-like volva.
Ring: near apex of stem; white.
Gills: white; virtually free.
Smell: when old unpleasant, cheesy.
Spore-print: white.
Habitat: deciduous woodland, especially with beech and oak. Autumnal. Fairly common.
POISONOUS, often fatal.

The streaky greenish cap, the sac-like volva, presence of a ring and the white spore-print are diagnostic of this deadly fungus.

AMANITA CITRINA
False Death Cap

Syn: *A. mappa*
Cap: 6-8 cm diam.; convex then flat; pale cream to pale lemon-yellow ornamented with a few large, or several smaller, thick, flat, whitish to brownish patches of velar tissue.
Stem: 8-11 cm high, 1-2 cm wide, tall in proportion to cap; white, cylindrical with a conspicuous basal bulb, up to 3 cm wide, with margin indicated by a prominent rim representing the volva.
Ring: near apex; white.
Gills: white to very pale cream.
Smell: of raw potato.
Spore-print: white.
Habitat: deciduous and coniferous woodland. Autumnal. Very common.

Formerly confused with the Death Cap and regarded as poisonous, but in reality it is harmless. *A. citrina* is easily separated from *A. phalloides* by cap-colour, and lack of a sac-like volva.

AMANITA MUSCARIA
Fly Agaric

Cap: up to 15 cm diam., convex, flattened to saucer-shaped, scarlet ornamented with white warts which become more dispersed toward the striate margin. With age the warts gradually disappear and may be completely lacking. The cap colour often fades to reddish-orange. In the button stage the cap is covered by thick white velar tissue which eventually cracks to form the scales and expose more and more of the red surface as expansion occurs.
Stem: up to 20 cm high, 3 cm wide, white to pale-yellowish above, cylindrical and brittle with a slightly broader base surmounted by a series of concentric zones of white scales representing the volva.
Ring: near apex of stem, white to very pale-yellowish.
Gills: white, almost free.
Habitat: Under birch, occasionally with pine and other trees. Autumnal. Common. POISONOUS but seldom lethal.

AMANITA PANTHERINA

Cap: 6-8 cm diam., convex, then flat, dark greyish-brown to olive-brown, sometimes paler and more yellowish-brown, ornamented with numerous, uniformly distributed small white pyramidal warts; margin somewhat striate.
Stem: 7-10 cm high, 1-1·5 cm wide, white, rather tall, cylindrical, slightly enlarged below where the volva disrupts into concentric rings, the uppermost forming a narrow, close-fitting but free collar or ridge.
Ring: near middle of stem, white.
Gills: white, free.
Flesh: white becoming orange-yellow under cap cuticle with potassium hydroxide.
Spore-print: white.
Habitat: deciduous woodland. Autumnal. Rare. POISONOUS.

Recognized by the brown cap ornamented with numerous small contrasting white pyramidal scales, the striate cap-margin, and collar-like formation of the volva.

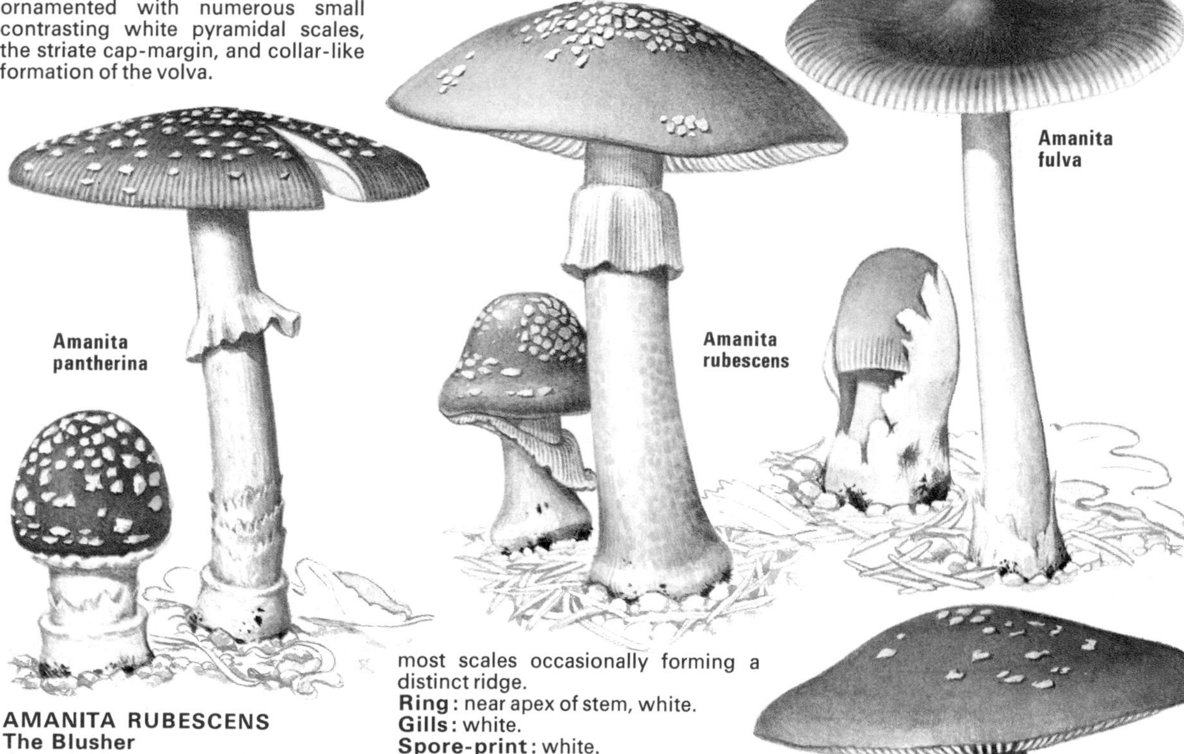

AMANITA RUBESCENS
The Blusher

Cap: 8-12 cm diam., at first strongly bell-shaped, then convex, finally flat to shallowly saucer-shaped, varying in colour from pale pinkish-brown with a darker red-brown centre to entirely red-brown, ornamented with thin mealy or hoary patches of whitish, greyish or pale volval tissue which tend to disappear with age.
Stem: 8-11 cm high, 2·5-3·5 cm wide, stout and stocky, cylindrical with somewhat enlarged base, whitish becoming pale pinkish-brown below, especially when handled. Volva reduced to inconspicuous rows of concentric scales.
Ring: near apex of stem; white.
Gills: white, often spotted red with age.
Flesh: white becoming pinkish when cut and in insect holes.
Spore-print: white.
Habitat: deciduous and coniferous woodland. Late summer and autumn. One of the first mushrooms to fruit. Very common. Edible, but best avoided owing to possible confusion with poisonous species.

AMANITA EXCELSA
Syn. *A. spissa*

Cap: 9-12 cm, convex, flat, then saucer-shaped, greyish to umber-brown ornamented with thin, greyish mealy to hoary patches of velar tissue which may eventually disappear; margin smooth.
Stem: 8-10 cm high, 2-2·5 cm wide, often rather stocky, white, cylindrical, base bulbous with concentric rings of scales representing the volva, uppermost scales occasionally forming a distinct ridge.
Ring: near apex of stem, white.
Gills: white.
Spore-print: white.
Habitat: deciduous and coniferous woodland. Autumnal. Fairly common.

A. pantherina is similar but differs in the white pyramidal warts on the cap (which has a striate margin) and the ridge-like rim of volval warts at the base of the stem. *A. rubescens* differs in colour and in the reddening of the stem both internally and externally.

Amanita excelsa

AMANITA FULVA
Tawny Grisette
Syn. *Amanitopsis fulva*

Cap: 3·5-5 cm diam., long remaining acorn-shaped, then bell-shaped, finally flat but often with central boss, bright tawny-brown, often darker at centre, naked. Margin striate, fluted or grooved.
Stem: up to 11 cm high, 1 cm wide, tall, fragile, hollow, cylindrical, pale-tawny with well-developed, sac-shaped volva, same colour as stem, and which often clings to the slightly enlarged base.
Ring: absent.
Gills: white, free.
Spore-print: white.
Habitat: coniferous and deciduous woodland. From late summer to autumn; one of the earliest mushrooms to appear. Very common. Edible.

A. crocea, a closely related but much rarer species, is more robust with a brighter orange-brown cap, and with a tendency for the entire stem surface to disrupt into coarse scales.

LEPIOTA

A genus of about 60 fungi, mostly rather small, delicate species with caps 1-5 cm diam., but a few, often placed in the genus *Macrolepiota*, are amongst our largest mushrooms. Most species have a scaly cap formed by the disruption of the surface. Most have a well-defined ring or ring zone, but this is sometimes deciduous, and only visible on young specimens. All species lack a volva, although there may be conspicuous scales at the base of the stem. The gills are free. Spore-print white. Most of the small species are toxic.

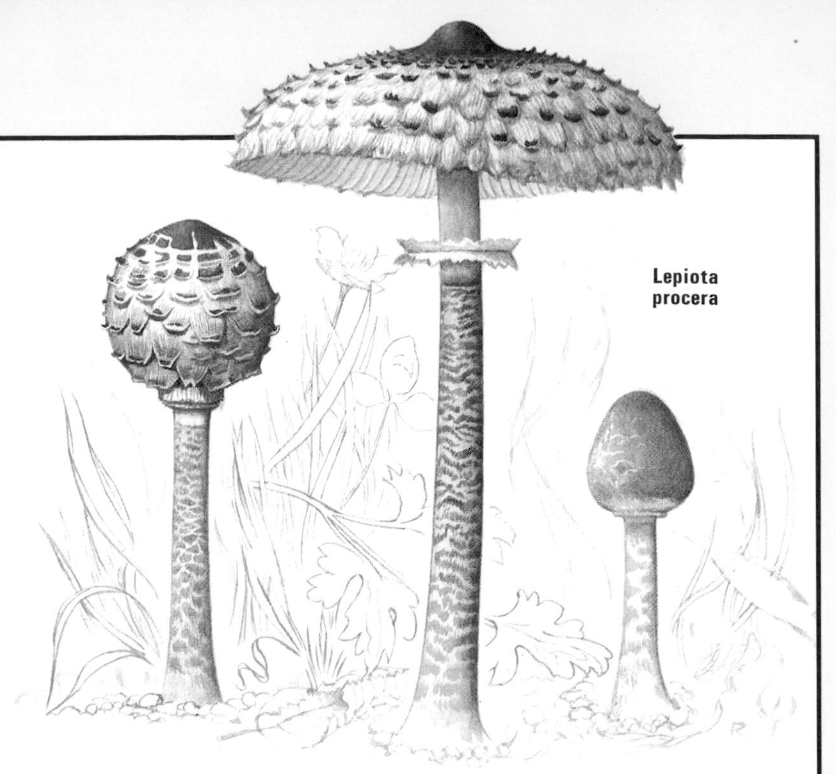

Lepiota procera

LEPIOTA CRISTATA

Cap: 2-4 cm diam., broadly campanulate, surface disrupting into tiny red-brown scales on a white background; the scales rapidly disappear except around a small, central, similarly coloured disc.
Stem: 2-5 cm high, 4-6 mm wide, white, cylindrical to slightly enlarged below.
Ring: near apex of stem, white, membranous, deciduous.
Gills: white, free.
Smell: unpleasant, sour, reminiscent of Earth Balls (*Scleroderma spp*).
Spore-print: white.
Habitat: deciduous woodland, often in grass along rides. Autumnal. Common. POISONOUS.

Easily confused with other small *Lepiota* species, but when examined under a strong lens most of these show minute pyramidal tufts of hairs over the central disc of the cap: these are lacking in *L. cristata.*

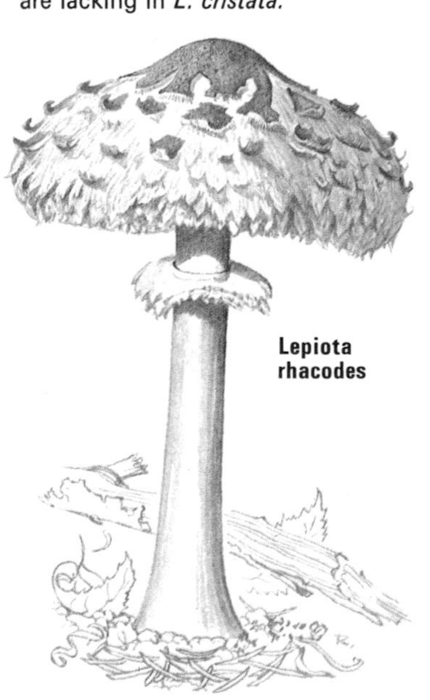

Lepiota rhacodes

LEPIOTA PROCERA
Parasol Mushroom

Cap: 12-22 cm diam., umbrella-shaped with nipple-like boss at centre, surface disrupting into large, often upturned, dark-brown scales on a dirty white, coarsely fibrillose background; scales larger and widely dispersed towards margin, smaller, more densely arranged at centre with nipple uniformly dark-brown.
Stem: 20-26 cm high, 1·5-2·0 cm wide, tall, narrow, cylindrical with bulbous base up to 4 cm wide, pale, ornamented with dark-brown, zig-zag markings.
Ring: large, spreading with double edge, whitish, movable.
Gills: soft to touch, free, attached to collar-like rim surrounding the stem apex.
Flesh: white.
Spore-print: white.
Habitat: pastures, edge of woods grassy rides. Aumtunal, Occasional. Edible.

Unmistakable, distinguished by habitat, dark scaly cap, snake-like markings on stem, movable ring and unchanging white flesh. The movable ring is a feature of the large *Lepiota* species. Another feature is that the cap is easily separated from the stem due to a 'ball and socket' attachment.

LEPIOTA RHACODES
Shaggy Parasol

Cap: 8-12 cm diam., convex, lacking a central boss, surface disrupting into yellowish-brown, upturned shaggy scales on a dirty-white fibrillose background, except at the centre which is uniformly coloured and often darker.
Stem: 12-15 cm high, 1·5-2·0 cm wide, whitish, cylindrical with bulbous base up to 3 cm wide.
Ring: large, spreading, whitish, movable.
Gills: deep, soft to touch, free, attached to collar-like rim surrounding stem apex.
Flesh: white, reddening in stem when cut.
Spore-print: white.
Habitat: deciduous and coniferous woodland and shrubberies. Autumnal. Fairly common. Edible.

Distinguished from all other large *Lepiota* species by its reddening flesh, and additionally from *L. procera* in stockier appearance, differently shaped cap with coarse, shaggy, yellow-brown scales and lack of snake-like markings on the stem. It is also distinguished by its preference for growing in woodland rather than pastures.

Lepiota cristata

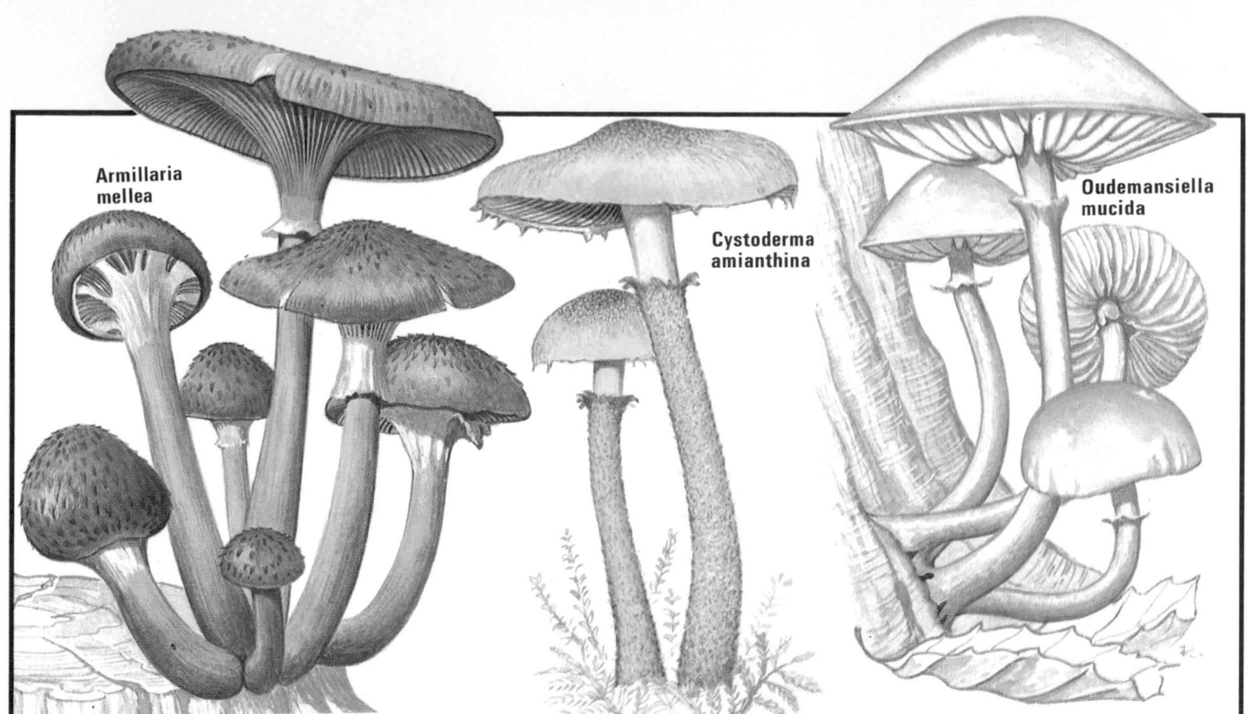

ARMILLARIA

Two species. Fruitbodies densely tufted at base of trunks. Parasitic.

ARMILLARIA MELLEA
Honey Fungus
Cap: 6-12 cm diam., convex, then flat to saucer-shaped, tan, tawny or cinnamon-brown, paler towards the indistinctly striate margin; ornamented with delicate, dark-brown, hair-like scales, prominent and crowded in young specimens, eventually disappearing except at centre in old fruitbodies.
Stem: 9-14 cm high, 1-2 cm wide, tough, cylindrical, pale-tawny, whitish at apex.
Ring: near apex of stem, thick, cottony, whitish, often with yellow flocci at margin.
Gills: dirty-whitish to flesh-coloured; adnate.
Flesh: white, soft. Taste acrid.
Spore-print: cream.
Habitat: at base of living and dead trunks or stumps. Autumnal. Very common. Edible when young.

CYSTODERMA

A genus of about 6 species. Fruitbody medium to small, caps yellow-ochre, orange-brown, reddish-brown or pale flesh-coloured with granular-mealy surface. Stems similarly coloured. Spore-print white.

CYSTODERMA AMIANTHINA
Syn: *Lepiota amianthina*
Cap: 2-3 cm diam., broadly bell-shaped with central boss, yellowish to ochre-brown, with granular mealy surface.
Stem: 4-6 cm high, 4-6 mm wide. same colour as cap, often darker at base, covered with mealy granules below ring, apex pale.
Ring: sometimes indistinct, same colour as cap, granular beneath.
Gills: adnate, cream.
Flesh: yellowish.
Spore-print: white.
Habitat: Heathy places or coniferous woodland, amongst grass and moss. Autumnal. Fairly common.

OUDEMANSIELLA

Four species. Cap viscid to glutinous but velvety in *O. longipes* and *O. badia*. Gills deep, distant.

OUDEMANSIELLA MUCIDA
Poached Egg Fungus
Syn: *Armillaria mucida; Collybia mucida*
Cap: 3-7 cm diam., soft, flabby, convex, very glutinous, white becoming flushed with grey, margin striate, almost translucent.
Stem: 5-7 cm high, 4-6 mm wide, often curved, tough, cartilaginous, cylindrical; often expanded disc-like at point of attachment; white to greyish.
Ring: spreading, white above, greyish below especially toward edge.
Gills: distant; deep, soft, white.
Spore-print: white.
Habitat: confined to beech, occurring in small clusters on trunks, branches, or on stumps. Autumnal. Common.

OUDEMANSIELLA RADICATA
Rooting Shank; Long Root
Syn: *Collybia radicata*
Cap: 5-8 cm., tough, convex, then flat, often coarsely radiately wrinkled around a central boss, sticky when wet, tacky when dry, varying in colour from pale buff or fawn to yellowish- or reddish-brown.
Stem: 10-18 cm high, 6-10 mm wide; tall, tough, cartilaginous, narrowly cylindrical, paler than cap, often greyish, longitudinally striate and usually twisted, prolonged below into a long root up to 30 cm in length.
Gills: distant, deep, white.
Spore-print: white.
Habitat: deciduous woodland, apparently terrestrial but arising from buried wood and roots etc. Late summer to autumn. Common.

Easily recognized by its sticky cap and tall elegant stem with long rooting base. *O. longipes* and *O. badia* both have a dry velvety brown cap and stem but otherwise resemble *O. radicata* in stature and habit.

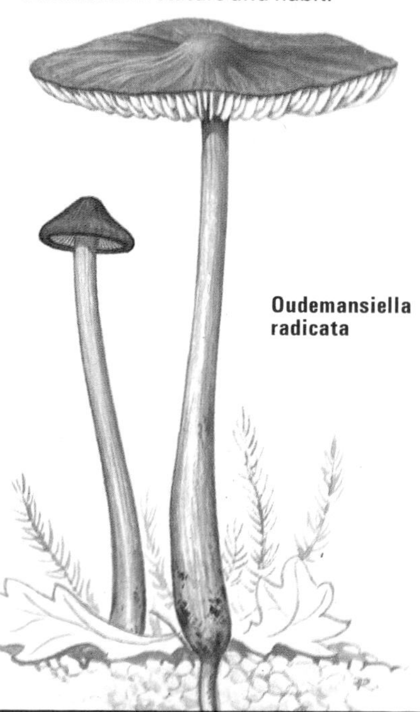

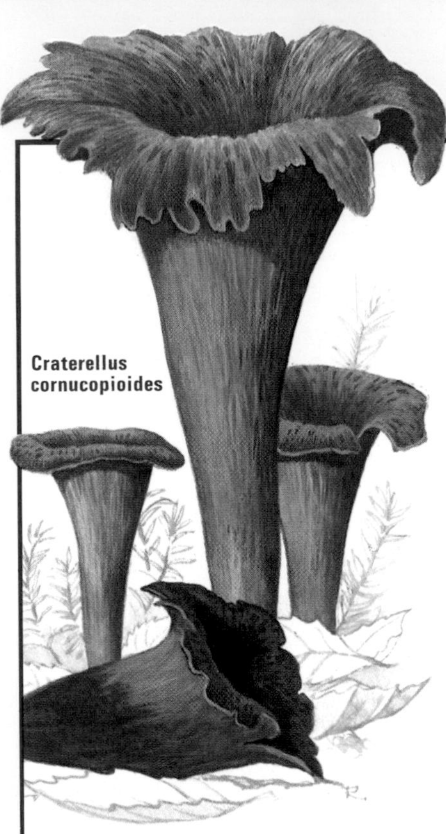

CRATERELLUS

A single species. Cap tubular with flared mouth, blackish.

CRATERELLUS CORNUCOPIOIDES
Horn of Plenty; Trumpet of Death
Cap: 2·5-7 cm high, tough, tubular, horn-shaped with flaring mouth; inner surface felty, sooty-brown to blackish, drying out to pale grey-brown; outer fertile surface smooth to slightly uneven, dark slate-grey, sometimes flushed ochre-coloured, with ash-grey hoary appearance.
Habitat: amongst fallen leaves in deciduous woodland, especially beech. Autumnal. Occasional. Edible and good.

HYGROPHOROPSIS

Three species with several varieties. Cap small to medium-sized, funnel-shaped with white, yellowish-ochre or orange suede-like surface and incurved margin. Gills crowded, similarly coloured. regularly and repeatedly forked. Spore-print white.

HYGROPHOROPSIS AURANTIACA
False Chanterelle
Syn: *Cantharellus aurantiacus; Clitocybe aurantiaca.*
Cap: 3-6 cm diam., funnel-shaped with enrolled margin, surface suede-like, orange or yellowish-orange.
Stem: 2-4 cm high, 5-7 mm wide, same colour as cap, often brownish below when old.
Gills: decurrent, crowded, repeatedly forked, deep orange.
Smell: not distinctive.
Spore-print: white.
Habitat: coniferous woodland and heaths. Autumnal. Common. Edible but worthless.

Recognized by the orange, funnel-shaped fruitbody with strongly enrolled margin and the repeatedly forked crowded orange gills. Frequently mistaken for the genuine Chanterelle which is altogether more fleshy and top-shaped, with irregularly branched fold-like shallow gills It also has a smell of apricots.

CANTHARELLUS

Fruitbodies thin or fleshy, funnel-shaped with broad, shallow, irregularly branched gill-like folds. 'Gills' not sharp-edged like those of other mushrooms, but resembling veins or wrinkles. Spore-print white to cream.

CANTHARELLUS CIBARIUS
Chanterelle
Cap: 2·5-6 cm diam., top-shaped, often slightly depressed at centre, gradually narrowed below into the short stalk; smooth, moist, entire fungus bright egg-yellow.
Stem: 2-6 cm high, 6-16 mm wide, short and squat.
Gills: decurrent, blunt, irregularly branched, fold-like and interconnected.
Smell: pleasant (of apricots).
Spore-print: white.
Habitat: deciduous woodland, especially on sandy or clay banks amongst moss. Autumnal. Occasional. Edible. Much sought after and collected for sale in markets. Easily dried for use in cooking.

The fleshy, egg-yellow fruitbody with irregularly branched, shallow, fold-like gills and smell of apricots is diagnostic. The only species liable to confusion with it is *Hygrophoropsis aurantiaca.* Here the fruitbody is much less fleshy, is funnel-shaped, orange and has true crowded gills which fork repeatedly in a regular manner.

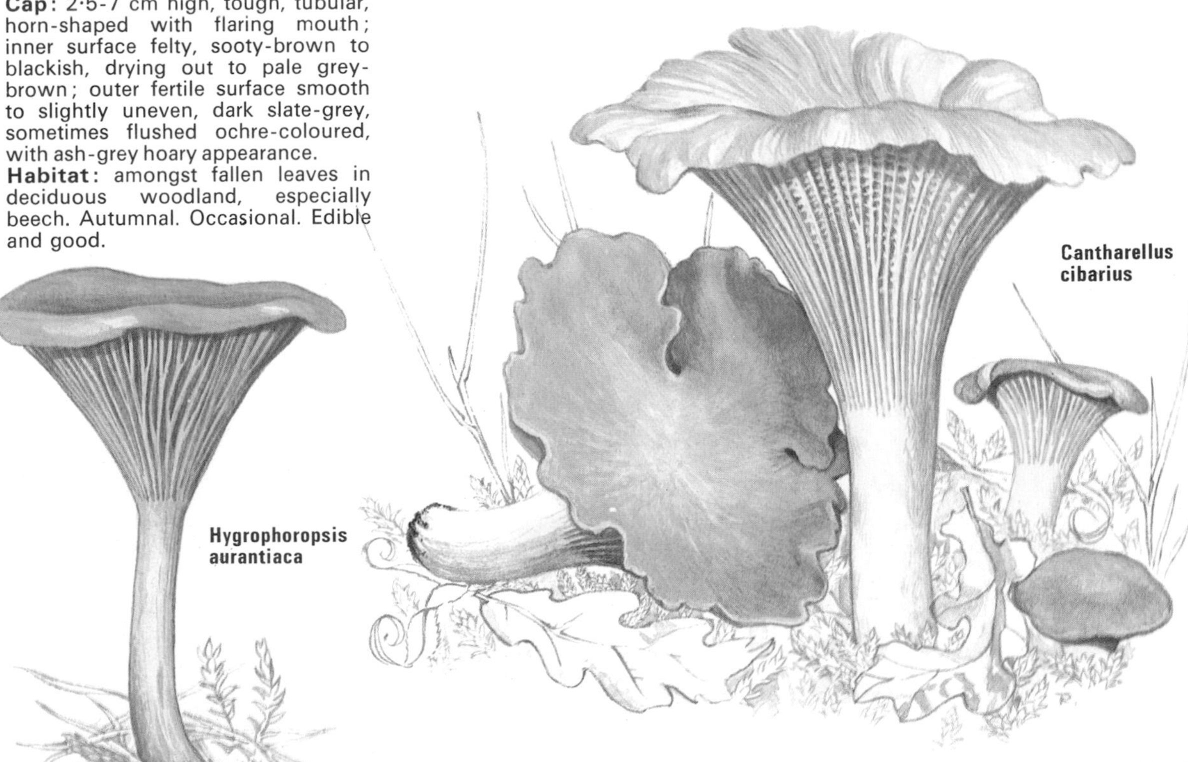

LACTARIUS
Milk Caps

A large genus of over 50 species. Cap small or robust, commonly funnel-shaped with dry or sticky, smooth or shaggy surface, variously coloured and frequently with conspicuous zones. Stem often narrowed below, sometimes pockmarked. Gills more or less decurrent, exuding a milky juice when cut or broken. In some species the milk is mild and tasteless in others exceedingly hot and peppery burning the tongue for some minutes after tasting – which it is quite safe to do. Smell often strong of curry powder or fenugreek in dried material.

LACTARIUS VELLEREUS

Cap: 8-16 cm diam., robust, funnel-shaped with enrolled margin, white, often patchily yellowish near edge, surface suede-like and rather shortly downy.
Stem: 4-6 cm high, 2-3 cm wide; short, squat, firm, whitish.
Gills: decurrent, distant, thick, white to pale-cream.
Milk: white, very hot and peppery.
Spore-print: white.
Habitat: Deciduous woodland. Autumnal. Occasional.

Readily recognized by large size, white cap with suede-like surface and white peppery milk.

L. piperatus differs in the very crowded gills and also in the less obvious and not at all downy suede-like texture of the cap surface.

Russula delica is a species which is also liable to confusion with this fungus but differs in having very brittle gills which fail to 'milk' when broken.

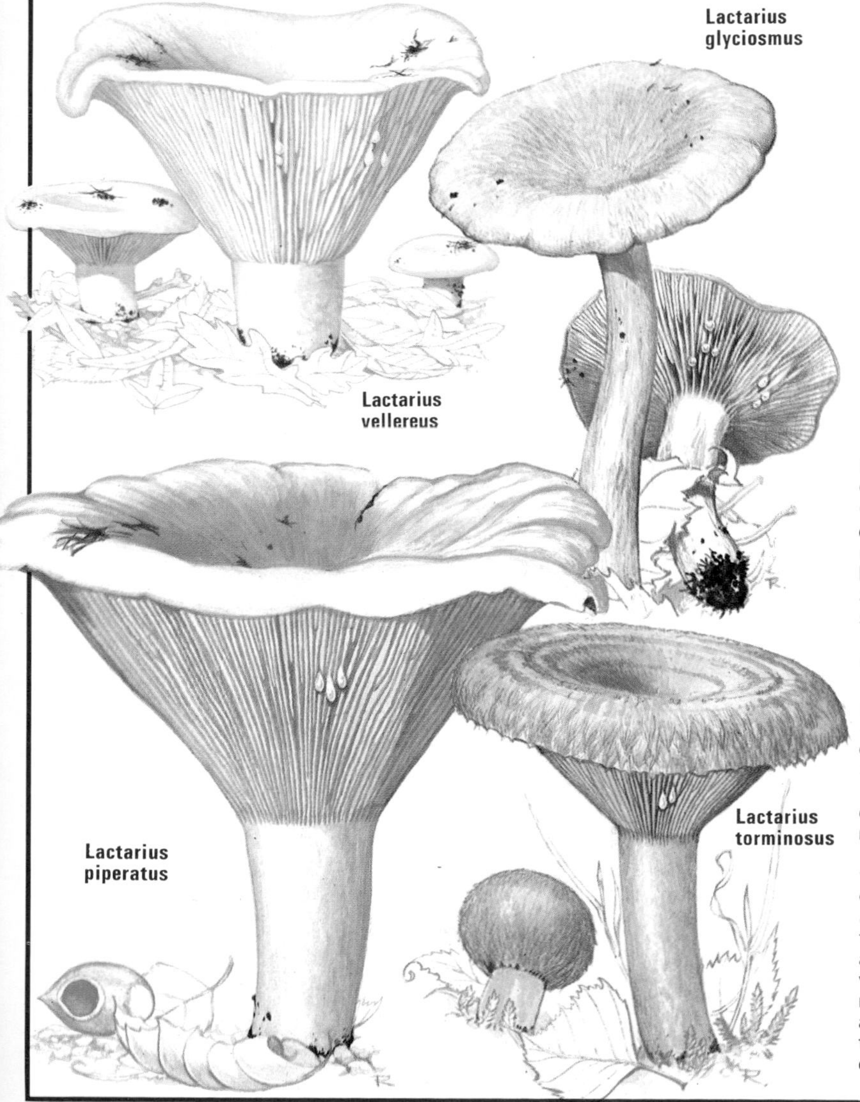

LACTARIUS TORMINOSUS

Cap: 4-9 cm diam., shallowly funnel-shaped, surface shaggy fibrillose, especially toward the enrolled margin, crushed strawberry pink with darker zones.
Stem: 5-6 cm high, 8-10 mm wide, pinkish flesh-coloured.
Gills: more or less decurrent, pale flesh-colour.
Milk: white, very peppery.
Spore-print: pale pinkish-buff.
Habitat: strictly with birch on heaths and commonland. Autumnal. Common.

The main points for recognition are the pink shaggy cap, the white peppery milk and the occurrence under birch.

A number of other species have shaggy caps and peppery milk. *L. pubescens* also grows with birch but the cap is whitish with a thick shaggy covering devoid of zonation. *L. mairei*, a species with orange-buff colouring, is associated with oak.

LACTARIUS PIPERATUS

Cap: 10-16 cm diam., funnel-shaped, with smooth white surface.
Stem: 4-6 cm high, 2-3 cm wide; short, stocky, white.
Gills: white, decurrent, shallow, very densely crowded.
Milk: white, plentiful, very peppery.
Spore-print: white.
Habitat: deciduous woodland. Autumnal. Occasional.

Fairly readily distinguished from *L. vellereus* by the smooth cap and densely crowded gills. *L. controversus* is a similar species but the cap is blotched red or pink and has pale, rosy-buff gills. It is found associated with poplars or with dwarf willows on sand-dunes. *L. glaucescens* is principally recognized by the milk slowly becoming faintly greenish-blue.

LACTARIUS GLYCIOSMUS
Coconut Smelling Milkcap

Cap: 2-6 cm diam., flat to slightly depressed, with a small central nipple, smooth, dry, pale greyish-lilac, sometimes dull-buff.
Stem: 4-5·5 cm high, 4-6 mm wide, same colour as cap but paler.
Gills: more or less decurrent, pinkish-buff to pale ochre.
Milk: white, mild becoming slightly peppery.
Smell: strong, sweetish, of desiccated coconut.
Spore-print: pale ochre-coloured.
Habitat: under birch on heaths and commonland. Autumnal. Fairly common.

The habitat, small greyish-lilac cap with central papilla and smell of coconut are the distinctive features. *L uvidus* has a tacky, violet-grey cap up to 10 cm diam., creamy to pinkish gills and a pale stem which becomes yellowish from below and finally marked with rusty spots. The bitter milk, at first white, becomes violet as does the cut flesh. This fungus grows in damp birch woods.

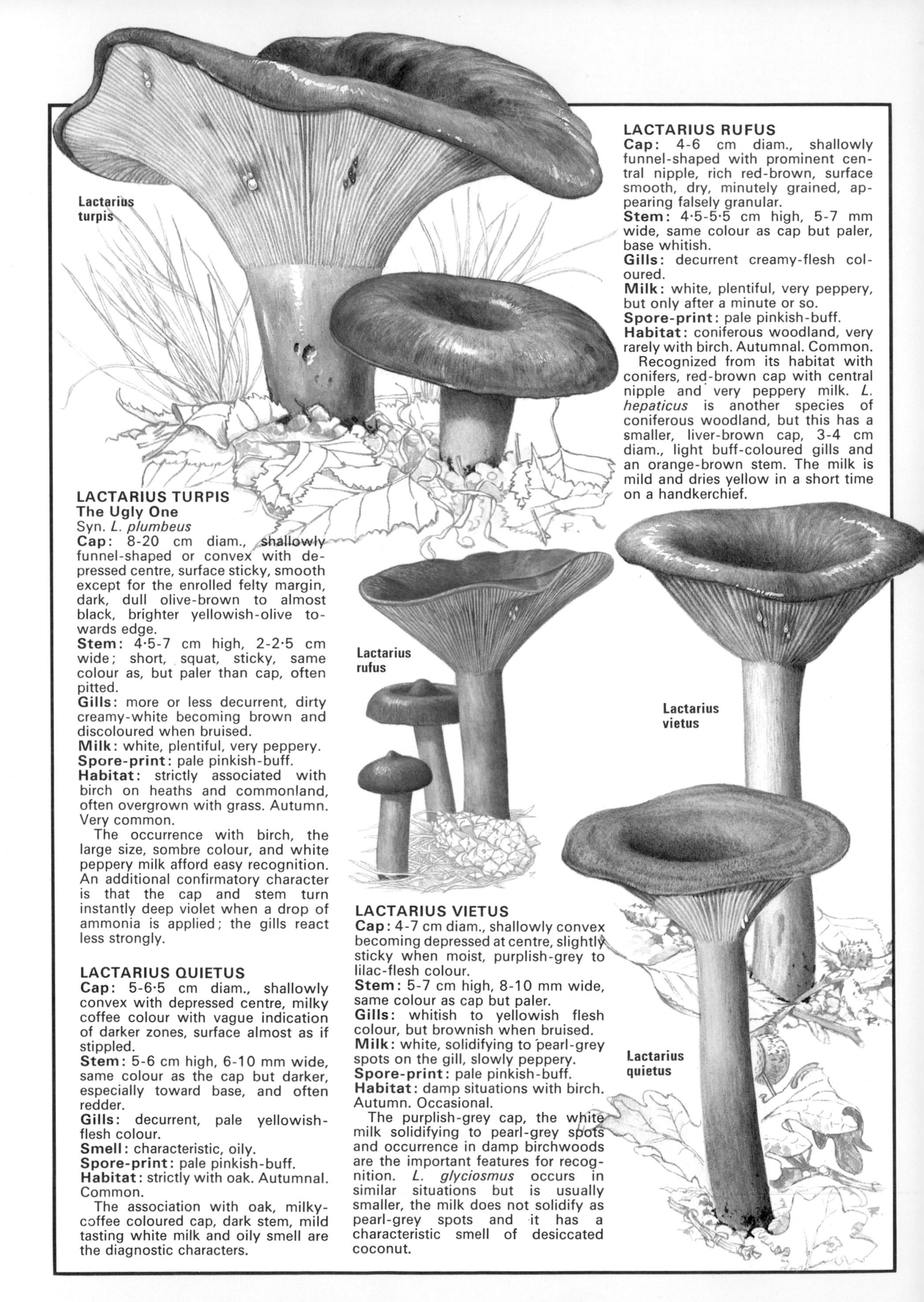

LACTARIUS RUFUS

Cap: 4-6 cm diam., shallowly funnel-shaped with prominent central nipple, rich red-brown, surface smooth, dry, minutely grained, appearing falsely granular.
Stem: 4·5-5·5 cm high, 5-7 mm wide, same colour as cap but paler, base whitish.
Gills: decurrent creamy-flesh coloured.
Milk: white, plentiful, very peppery, but only after a minute or so.
Spore-print: pale pinkish-buff.
Habitat: coniferous woodland, very rarely with birch. Autumnal. Common.

Recognized from its habitat with conifers, red-brown cap with central nipple and very peppery milk. *L. hepaticus* is another species of coniferous woodland, but this has a smaller, liver-brown cap, 3-4 cm diam., light buff-coloured gills and an orange-brown stem. The milk is mild and dries yellow in a short time on a handkerchief.

LACTARIUS TURPIS
The Ugly One
Syn. *L. plumbeus*

Cap: 8-20 cm diam., shallowly funnel-shaped or convex with depressed centre, surface sticky, smooth except for the enrolled felty margin, dark, dull olive-brown to almost black, brighter yellowish-olive towards edge.
Stem: 4·5-7 cm high, 2-2·5 cm wide; short, squat, sticky, same colour as, but paler than cap, often pitted.
Gills: more or less decurrent, dirty creamy-white becoming brown and discoloured when bruised.
Milk: white, plentiful, very peppery.
Spore-print: pale pinkish-buff.
Habitat: strictly associated with birch on heaths and commonland, often overgrown with grass. Autumn. Very common.

The occurrence with birch, the large size, sombre colour, and white peppery milk afford easy recognition. An additional confirmatory character is that the cap and stem turn instantly deep violet when a drop of ammonia is applied; the gills react less strongly.

LACTARIUS QUIETUS

Cap: 5-6·5 cm diam., shallowly convex with depressed centre, milky coffee colour with vague indication of darker zones, surface almost as if stippled.
Stem: 5-6 cm high, 6-10 mm wide, same colour as the cap but darker, especially toward base, and often redder.
Gills: decurrent, pale yellowish-flesh colour.
Smell: characteristic, oily.
Spore-print: pale pinkish-buff.
Habitat: strictly with oak. Autumnal. Common.

The association with oak, milky-coffee coloured cap, dark stem, mild tasting white milk and oily smell are the diagnostic characters.

LACTARIUS VIETUS

Cap: 4-7 cm diam., shallowly convex becoming depressed at centre, slightly sticky when moist, purplish-grey to lilac-flesh colour.
Stem: 5-7 cm high, 8-10 mm wide, same colour as cap but paler.
Gills: whitish to yellowish flesh colour, but brownish when bruised.
Milk: white, solidifying to pearl-grey spots on the gill, slowly peppery.
Spore-print: pale pinkish-buff.
Habitat: damp situations with birch. Autumn. Occasional.

The purplish-grey cap, the white milk solidifying to pearl-grey spots and occurrence in damp birchwoods are the important features for recognition. *L. glyciosmus* occurs in similar situations but is usually smaller, the milk does not solidify as pearl-grey spots and it has a characteristic smell of desiccated coconut.

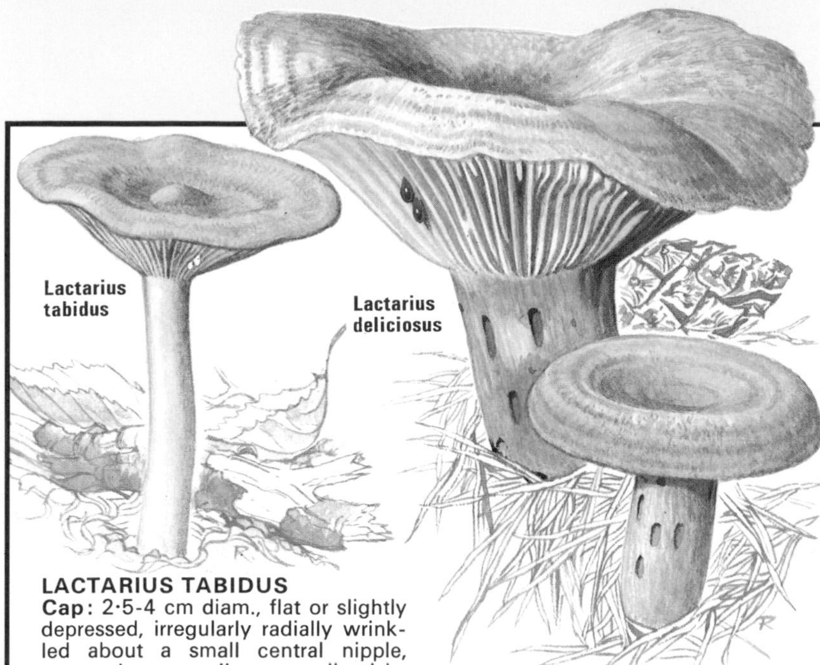

LACTARIUS TABIDUS

Cap: 2·5-4 cm diam., flat or slightly depressed, irregularly radially wrinkled about a small central nipple, orange-brown paling to yellowish-buff when dry, margin somewhat striate when moist.
Stem: 2-4 cm high, 4-6 mm wide, same colour as cap.
Gills: decurrent, buff.
Milk: mild, white, changing to yellow if allowed to dry on a handkerchief.
Spore-print: pale buff.
Habitat: deciduous woodland. Autumnal. Common.

The small orange-brown cap and mild milk drying yellow on a handkerchief are the main diagnostic characters of *L. tabidus.* There are several similar species with yellowing milk. *L. hepaticus,* a common pinewood species has a liver-coloured cap; *L. britannicus* is a slightly larger fungus of beechwoods with cap ranging from dark bay-brown around the papillate centre, through bright tawny-chestnut to creamy-orange at the margin.

LACTARIUS DELICIOSUS

Cap: 6-10 cm diam., convex with depressed centre or shallowly funnel-shaped, moist, reddish-orange with darker greenish zones and variable development of green staining, surface appearing almost as if granular-stippled.
Stem: 6-8 cm high, 1·5-2·0 cm wide, orange, often pitted.
Gills: orange-yellow, staining greenish.
Milk: brilliant carrot-colour eventually becoming wine-red in 30 minutes.
Spore-print: pale pinkish-buff.
Habitat: coniferous woodland, especially pine. Autumnal. Common. Edible.

An unmistakable species due to its orange colour and green staining of all parts, and the presence of vivid carrot-coloured milk.

Other species of *Lactarius* with brightly coloured milk include *L. chrysorrheus,* a species associated with oak in which the milk becomes quickly yellow on the gill.

CLITOCYBE

About 60 species ranging from large and robust to medium-sized fungi, mostly dull coloured and either top-shaped or funnel-shaped, with somewhat decurrent to strongly decurrent gills. Spore-print white, cream or even with pinkish flush. Species of *Omphalina* are very similar, with decurrent gills, but are smaller.

CLITOCYBE NEBULARIS
The Clouded Clitocybe

Cap: 7-16 cm diam., robust, fleshy, convex with a low central hump to more or less flat, cloudy-grey sometimes with a brown tinge, surface dry appearing to have a faint hoary bloom.
Stem: 7-10 cm high, 1·5-2·5 cm wide, cylindrical, same colour as cap but paler, often striate.
Gills: decurrent, crowded, dirty creamy-white.
Smell: characteristic, unpleasant.
Spore-print: creamy-white.
Habitat: deciduous woodland, especially in areas rich in humus, near piles of rotting leaves or grass, usually gregarious and sometimes forming fairy-rings. Autumnal. Common. POISONOUS.

Recognized by its cloudy-grey colour and characteristic smell.

CLITOCYBE ODORA
The Sea Green Clitocybe

Cap: 3-5 cm diam., flat to slightly depressed at centre, entire fungus beautifully blue-green.
Stem: 3·5-4·5 cm high, 6-8 mm wide.
Gills: slightly decurrent, paler than cap.
Smell: strong, sweet, fragrant, of aniseed.
Spore-print: white.
Habitat: deciduous woodland, often gregarious in small groups. Autumnal. Occasional.

Readily identified by its sea-green colour and strong fragrant smell.

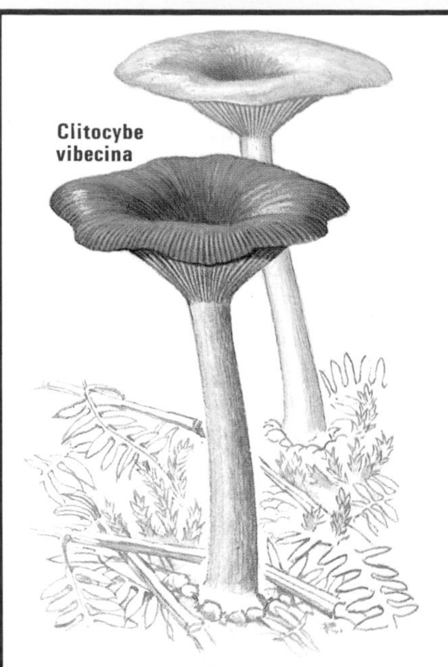

CLITOCYBE VIBECINA
Cap: 2·5-4 cm diam., convex with depressed centre to funnel-shaped, when moist uniformly watery grey-brown with striate margin, but drying to a pale opaque biscuit colour with a dark brown spot in the base of the funnel.
Stem: 3-4 cm high, 5-6 mm wide, greyish with white woolly base.
Gills: decurrent, pale grey-brown.
Smell: of meal when crushed.
Spore-print: white.
Habitat: deciduous and coniferous woodland, also heathland under bracken. Autumnal but persisting well into winter. Common.

There are several closely-related, small, grey-brown species of which *C. vibecina* is the most frequent. An important feature of note is the completely different appearance of the moist and dry states of these fungi.

C. dicolor differs in lack of mealy smell and in the stem being proportionally longer and distinctly darker at the base. *C. phyllophila* has a lead-white cap up to 7 cm diam., which soon discolours, becoming creamy-coloured.

CLITOCYBE FLACCIDA
Syn: *Clitocybe inversa*
Cap: 4-7 cm diam., tough, elastic, funnel-shaped with enrolled margin, surface smooth, tan, becoming somewhat reddish brown when old but paling when dry.
Stem: 4-6 cm high, 5-7 mm wide, tough, same colour as cap, base woolly.
Gills: decurrent, crowded, pale to yellowish.
Spore-print: white to pale cream.
Habitat: deciduous woodland, often gregarious or clustered sometimes in fairy-rings. Autumnal but persisting well into winter. Fairly common.

A similar fungus of coniferous woods is sometimes distinguished as *C. gilva* (Syn: *C. splendens*) and has an altogether paler yellowish cap. Both species have tiny spherical roughened spores.

C. infundibuliformis is another tough-elastic species of deciduous woodland. Flesh pale to pale tan-coloured, funnel-shaped cap, 4-5 cm diam., a firm, rather narrow stem, similarly coloured, about 5 mm wide and whitish decurrent gills.

RUSSULA
Brittle Gills

Probably the largest genus with over 120 species. Difficult to identify, it is essential to taste each specimen to note whether it is hot or mild, and also to take a spore-print on glass to match the exact shade somewhere between pure white and ochre. The species, often brilliantly coloured in shades of pink, red, purple, orange, yellow, green etc, range from small to robust and have caps which are mostly shallowly convex with shortish fragile stems. The easiest generic character to recognize is the brittle nature of the gills when the finger is lightly run over them. Further, with few exceptions (*R. nigricans* and relatives), the gills are all of the same length, practically without intervening short gills.

RUSSULA NIGRICANS
Cap: 9-15 cm diam., convex then depressed, pale then rapidly patchily brown becoming uniformly dark-brown, finally almost black.
Stem: 5-7 cm high, 2-3 cm wide, becoming brown.
Gills: thick, adnate, exceptionally distant, extremely brittle with intervening short gills, dirty creamy-white, red where bruised.
Flesh: firm, white but reddening and finally blackening when broken.
Habitat: deciduous woodland, especially beech. Autumnal. Common.

The sombre colour and exceptionally distant gills with intervening short gills make this an easy species to recognize. It often persists in a completely black rotten condition for many months and may be parasitized by one or other of two small whitish or greyish agarics: *Nyctalis parasitica* or *Nyctalis asterophora* (see p. 24). *Russula acrifolia* and *R. adusta* are similar, but have crowded, subdecurrent gills and intervening short gills.

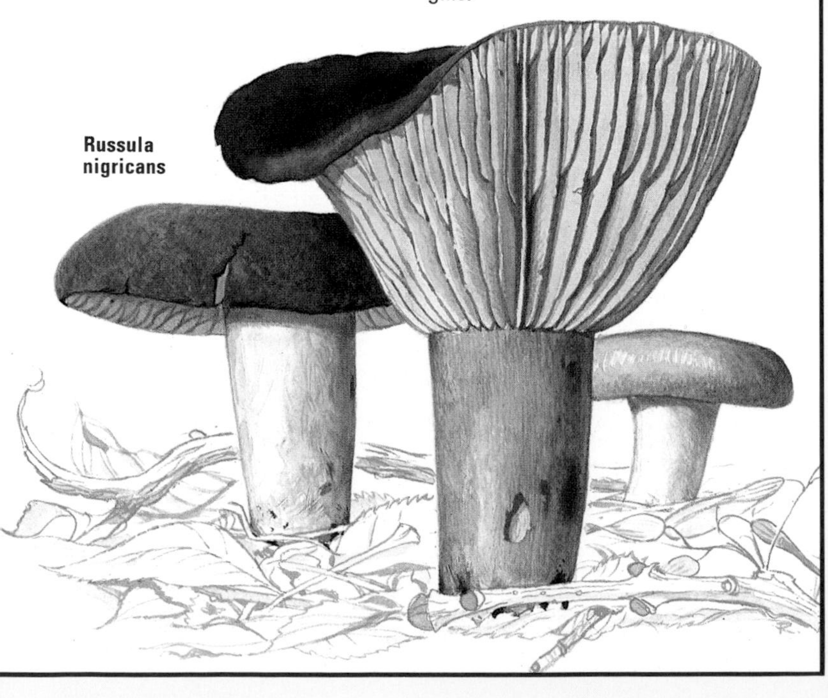

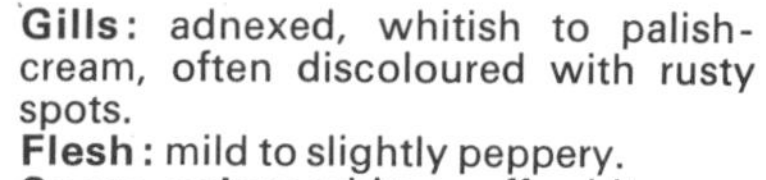

Russula
sardonia

RUSSULA EMETICA

The Sickener

Cap: 4-7 cm diam., convex then depressed, brilliant scarlet with a moist shiny surface, margin eventually coarsely striate.
Stem: 5-8 cm high, 1·5-2 cm wide, rather tall, fragile, pure white, the lower portion usually somewhat club-shaped.
Gills: adnexed, white.
Flesh: very hot.
Spore-print: pure white.
Habitat: coniferous woodland. Autumnal. Common. May cause sickness if eaten raw.

The habitat, brilliant colour, white spore-print, rather tall stature and hot peppery taste are the vital characters for correct identification. There are a number of closely related forms with slightly differing micro-characters. *R. mairei* is almost indistinguishable, except in growing with beech and having a shorter, firmer cylindrical stem.

RUSSULA SARDONIA

Syn: *Russula drimeia*
Cap: 6-10 cm diam., convex to broadly bell-shaped, varying from dark reddish- or violet-purple to purplish-black.
Stem: 7-12 cm high, 1·5-2 cm wide, rather tall, beautifully purple.
Gills: adnexed, primrose, often with watery droplets along the edge.
Flesh: white, very hot.
Spore-print: cream.
Habitat: coniferous woodland. Autumnal. Common.

R. queletii, another pinewood species, although much rarer, is virtually indistinguishable except for the whitish to very pale-cream gills. *R. caerulea* also occurs with pines and has a similarly coloured cap with shiny surface, although with a prominent central nipple, but the flesh is mild or only slightly peppery, the stem is white and the spore-print ochre. *R. atropurpurea* has a dark purplish-red to almost black cap but occurs in deciduous woodland. It has whitish gills, a somewhat peppery taste, and a white to off-white spore-print.

RUSSULA ATROPURPUREA

Cap: 4-10 cm diam., convex sometimes with slight hump, moist, dark reddish-purple to almost black, old specimens often mottled yellow at centre.
Stem: 5-7 cm high, 1·5-2 cm wide, short, squat, white, with rust-coloured base.
Gills: adnexed, whitish to palish-cream, often discoloured with rusty spots.
Flesh: mild to slightly peppery.
Spore-print: white to off-white.
Habitat: deciduous woodland. Autumnal. Common.

Amongst species of deciduous woods, *R. atropurpurea* is distinguished by its almost black cap and off-white spore-print. *R. fragilis* is similar but altogether smaller and more delicate with very peppery flesh and a white spore-print. The cap, seldom more than 4 cm diam., is patchily purplish pink at the margin with a dark depressed centre.

Dark purplish forms of *R. xerampelina* can be separated by having a smell of crab, a pale ochre spore-print and, a green reaction on the stem when rubbed with a crystal of Ferric Alum.

RUSSULA SANGUINEA

Cap: 7-10 cm diam., shallowly convex, then flat often with slight hump at centre, blood red, with a moist, almost granular look.
Stem: 8-12 cm high, 1·5-2 cm wide, whitish, flushed strongly pink.
Gills: slightly decurrent, crowded, ivory.
Flesh: white, hot.
Spore-print: cream.
Habitat: coniferous woodland, autumnal. Occasional.

The colour, strong pink flush on the stem, hot taste, cream spore-print and association with pines are the important points for recognition.

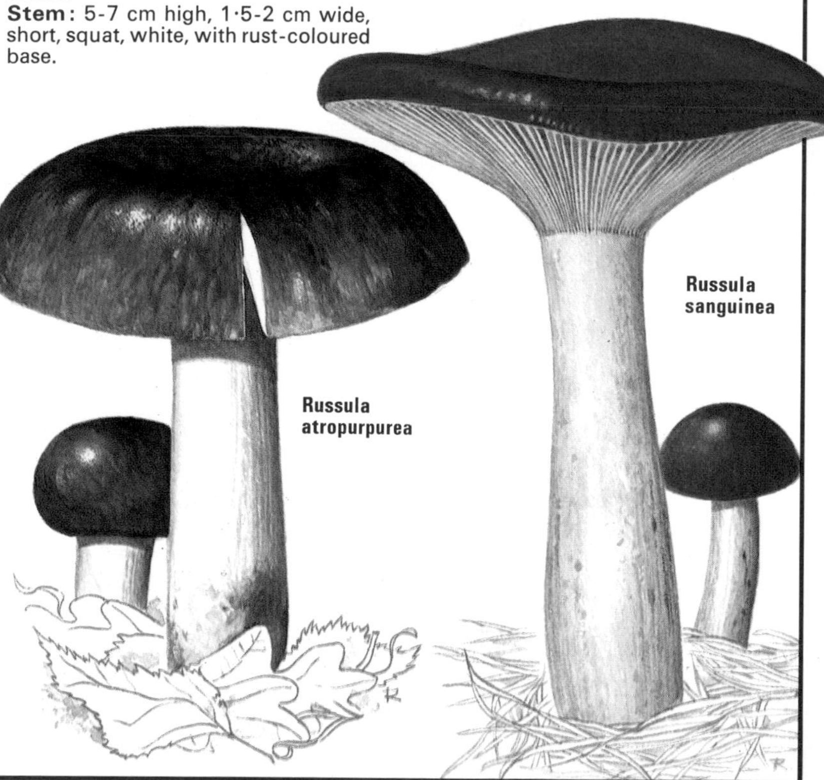

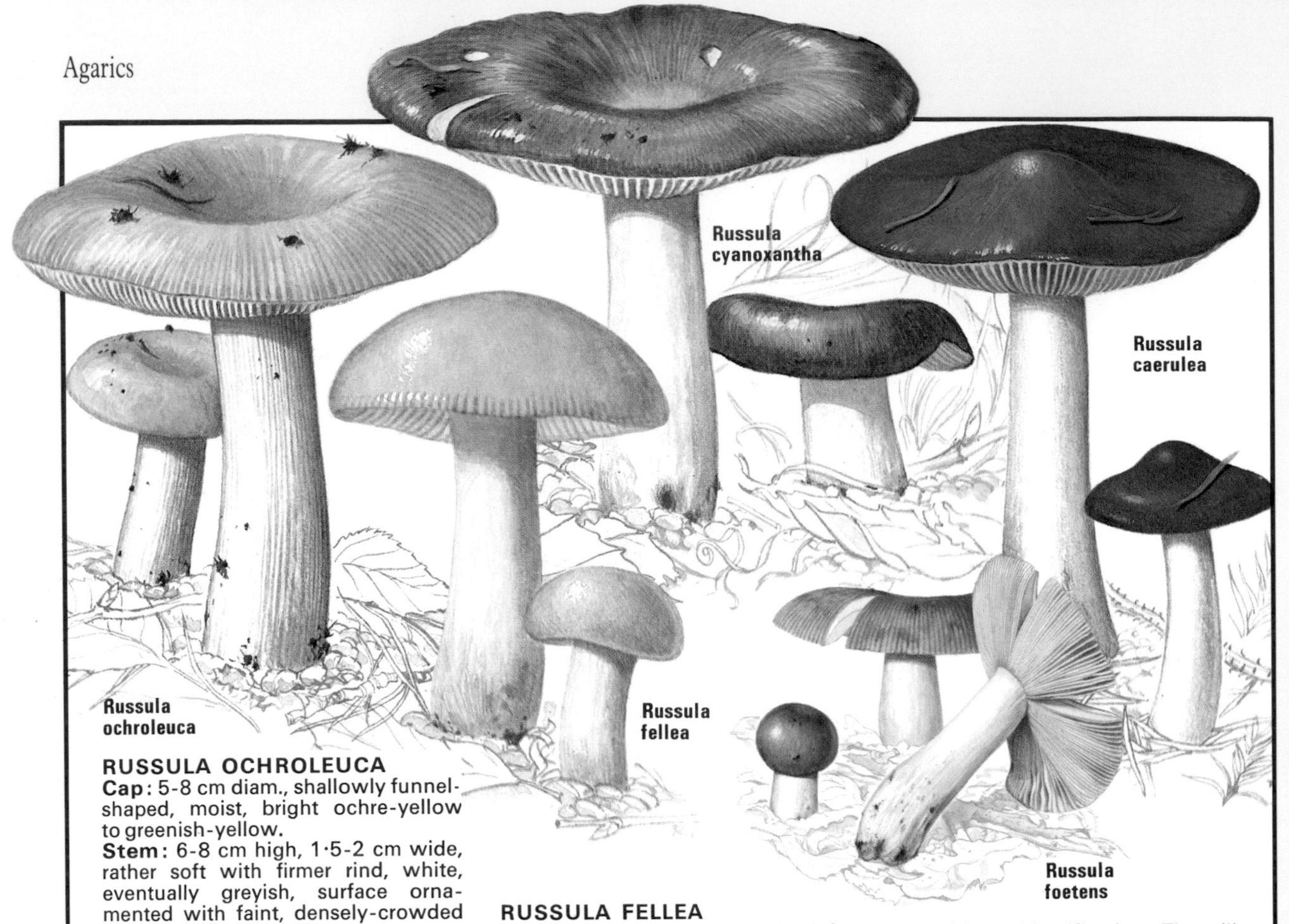

RUSSULA OCHROLEUCA
Cap: 5-8 cm diam., shallowly funnel-shaped, moist, bright ochre-yellow to greenish-yellow.
Stem: 6-8 cm high, 1·5-2 cm wide, rather soft with firmer rind, white, eventually greyish, surface ornamented with faint, densely-crowded short, raised longitudinal lines.
Gills: whitish.
Flesh: mild to moderately hot.
Spore-print: pale-cream.
Habitat: coniferous and deciduous woodland. Autumnal but persisting late in the season. Very common.

R. ochroleuca is one of the commonest Russulas and fairly easy to recognize although liable to confusion with *R. fellea.* The latter is quite frequent in beech woods and is usually smaller and more brownish in colour. The cap ranges from pale staw at the margin to deep staw at the centre. The fungus has a fruity smell resembling apples.

RUSSULA CAERULEA
Cap: 4-7 cm diam., convex with central nipple, shiny, uniformly dark purplish-violet.
Stem: 6-8 cm high, 1-1·5 cm wide, tall, white.
Gills: cream then yellow.
Flesh: mild.
Spore-print: ochre.
Habitat: coniferous woodland. Autumnal. Occasional.

The very dark cap with central nipple, together with yellow gills and ochre spore-print, white stem and coniferous habitat are the specific characters to note. *Russula sardonia* and *R. queletii,* both species of coniferous woods, have similarly coloured caps but they lack a central nipple, the gills and spore-print are much paler and the stems are pink. The taste is hot or very hot.

RUSSULA FELLEA
Cap: 4-5 cm diam., convex, varying from pale straw at the margin to darker straw at centre, surface with oiled or waxed appearance.
Stem: 4-6 cm high, 1·5-1·8 cm wide, same colour as cap but paler.
Gills: adnexed, same colour as cap.
Flesh: peppery, with a smell of apples or geraniums.
Spore-print: pale-cream.
Habitat: deciduous woodland, especially beech. Autumnal. Fairly common.

This species is similar to *R. ochroleuca* but is distinguished by the more brownish colour of the entire fruitbody and the smell of apples, as well as being more compact.

R. claroflava is a species associated with birch, often growing in sphagnum moss around margins of lakes and in swampy situations. It has a chrome-yellow cap, cream gills and the entire fungus including the flesh slowly blackens.

RUSSULA CYANOXANTHA
Cap: 7-10 cm diam., convex, slightly depressed at centre, moist, dark greyish-purple with olive tones.
Stem: 8-11 cm high, 1·5-2 cm wide, white, hard.
Gills: white, softly pliable.
Flesh: mild.
Spore-print: white.
Habitat: deciduous woodland. Autumnal. Common.

The purplish-grey cap with olive tones and elastic white gills are sure aids to identification. The gills are much less brittle than those of other *Russula* species.

Russula parazurea and *R. ionochlora* are similar but smaller species with a distinct hoary bloom to the cap and a cream spore-print. *R. heterophylla* and *R. aeruginea,* both associated with birch, have green caps; the former is rather rare and has a white print, the latter common with a buff-coloured print.

RUSSULA FOETENS
Cap: 9-14 cm diam., almost globular in the young stage, but later convex, slimy to glutinous, dingy ochre-brown, with conspicuously grooved margin.
Stem: 10-12 cm high, 2-4 cm wide, hollow, rather brittle, much paler than cap.
Gills: often weeping with watery droplets along edge, dingy white flushed with pale straw colour.
Flesh: peppery, with strong, foetid, oily smell.
Spore-print: cream.
Habitat: deciduous woodland. Autumnal. Fairly common.

The robust habit, dull brown cap with grooved margin, and the strong smell are diagnostic. *R. laurocerasi* differs in having a less strong smell of bitter almonds or crushed leaves of Cherry Laurel. *Russula sororia,* found under oaks, is a smaller, flatter species with greyish-brown cap and grooved, tuberculate margin.

FLAMMULINA

A single species characterized by lignicolous habit, bright tan-coloured, sticky cap, dark-brown velvety stem and white spore-print.

FLAMMULINA VELUTIPES
The Velvet Shank

Cap: 2·5-5 cm diam., shallowly convex becoming flat, bright-yellowish or orangey-tan, often darker and more brownish at the centre, moist becoming sticky when wet, shiny when dry.
Stem: 2·5-5 cm high, 4-6 mm wide, very dark-brown and conspicuously velvety, paling to yellow nearer the cap.
Gills: adnexed, rather distant, pale creamy-yellow.
Habitat: on trunks and branches, especially of dead elm, often in considerable quantity, forming small tiered clusters, very rarely growing from buried roots. Late autumn and winter.

TRICHOLOMOPSIS

Three lignicolous species. Fruitbodies tough or fleshy, often rather robust with distinctly or indistinctly felty surface, which may be brightly coloured, yellow, purple etc. Spore-print white.

TRICHOLOMOPSIS RUTILANS
Plums and Custard

Syn: *Tricholoma rutilans*
Cap: 6-12 cm diam., convex to broadly bell-shaped, yellow, densely covered with tiny fleck-like purple scales which are continuous at centre but nearer the margin become pulled further apart to show more and more of the yellow background.
Stem: 6-9 cm high, 1-1·5 cm wide, similar in colour to cap, likewise densely flecked below with purple scales which disappear towards the yellow apex.
Gills: yellow.
Flesh: yellowish.
Spore-print: white.
Habitat: on conifer stumps. Autumnal. Fairly common.

The combination of yellow and purple colours, and occurrence on conifer stumps makes this fungus easy to recognize.

T. platyphylla is a tough grey-brown fungus with large flabby cap 5-12 cm diam., which is rather streaky and slightly fibrillose; the gills are very deep, distant and whitish.

LEPISTA

Six species. Fruitbodies mostly fairly robust and fleshy and in some species bright lilac in parts. Gills more or less sinuate. Spore-print pale pinkish.

LEPISTA SAEVA
Blewit

Syn: *Tricholoma saevum; T. personatum; L. personata*
Cap: 6-8 cm diam., convex then flat, moist, varying from buff to greyish-buff.
Stem: 5-6 cm high, 1·5-2 cm wide, often enlarged at base, bright violet with streaky fibrillose surface.
Gills: whitish to pale flesh-coloured.
Spore-print: pale pinkish.
Habitat: in grassland, often forming fairy-rings. Autumnal. Uncommon. Edible and good, sometimes sold in shops.

Similar to *L. nuda* which differs in its woodland habitat and in the bright violet colour of all parts of the fruitbody.

LEPISTA NUDA
Wood Blewit

Syn: *Tricholoma nudum*
Cap: 6-10 cm diam., shallowly convex, becoming flat, often with greasy or water-soaked appearance, varying in colour from entirely violet to reddish-brown with violet tint localized to margin.
Stem: 6-8 cm high, 1·5-2 cm wide, bright bluish-lilac.
Gills: at first vivid violet becoming pinkish with age.
Spore-print: pale pinkish.
Habitat: deciduous woodland, compost heaps. Late autumn persisting into winter. Common. Edible and good, sometimes sold in shops.

Unlikely to be confused with other mushrooms provided a check is made on the colour of the spore-print. There are many *Cortinarius* species which show varying amounts of lilac or violet tints but these all have a rusty-brown print.

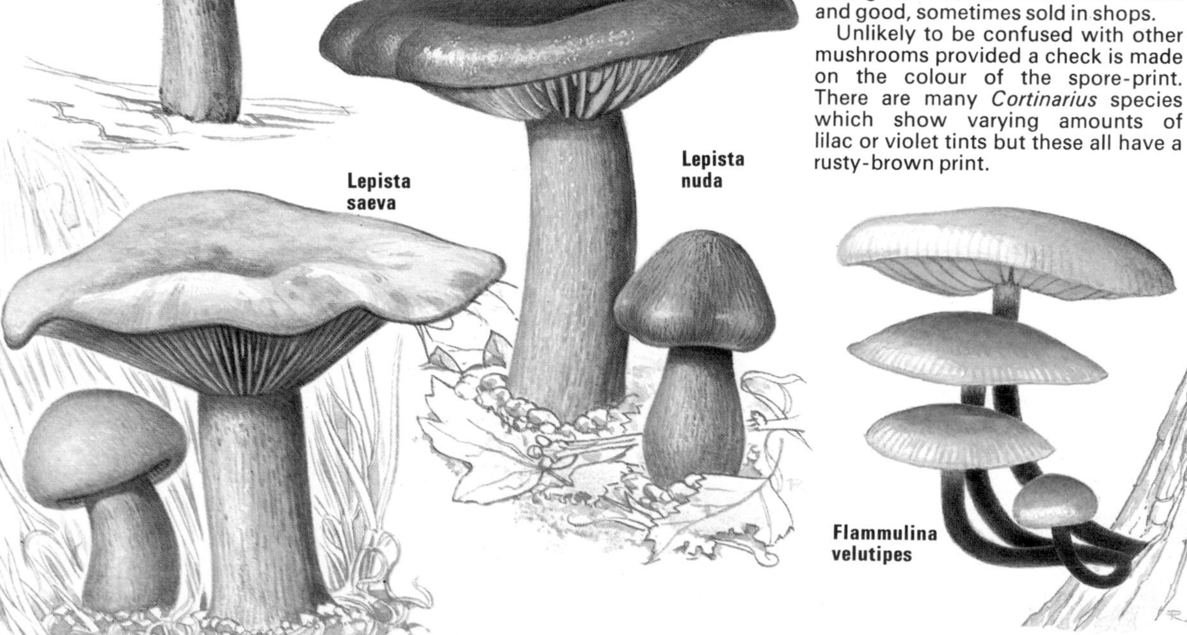

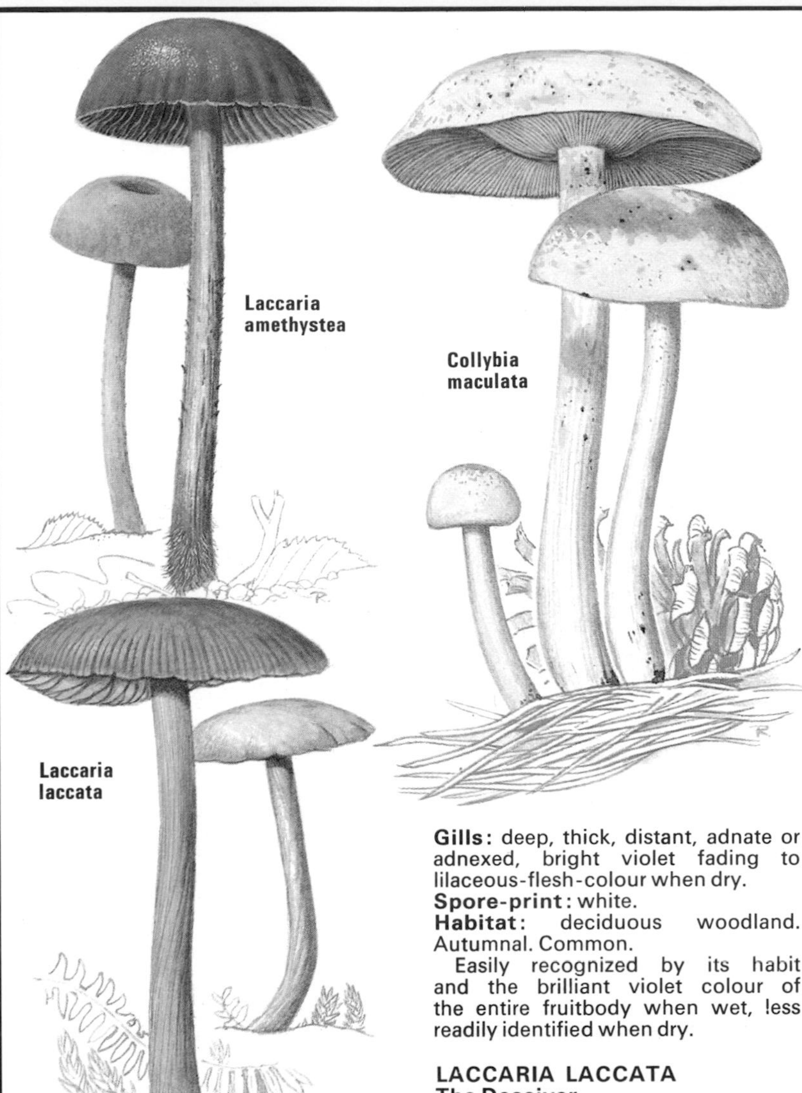

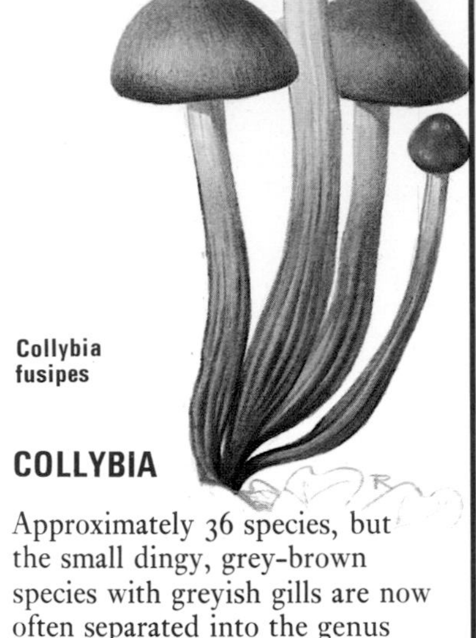

LACCARIA

Seven to eight species. Cap small to medium, often brightly coloured and usually with scurfy or felty surface which assumes a completely different aspect when dry to that when wet. Gills rather thick, distant, waxy and pinkish or deep amethyst. Spore-print white.

LACCARIA AMETHYSTEA
Amethyst Deceiver
Syn. *L. amethystina*
Cap: 2·5-4 cm diam., convex then shallowly convex, often depressed at centre, surface scurfy-felty, when moist entire fungus bright violet, but when dry this fades to pale buff with faint lilac tint.
Stem: 4-6 cm high, 6 mm wide, deep violet, but paler when dry.
Gills: deep, thick, distant, adnate or adnexed, bright violet fading to lilaceous-flesh-colour when dry.
Spore-print: white.
Habitat: deciduous woodland. Autumnal. Common.

Easily recognized by its habit and the brilliant violet colour of the entire fruitbody when wet, less readily identified when dry.

LACCARIA LACCATA
The Deceiver
Cap: 2-4 cm, shallowly convex to broadly bell-shaped, sometimes slightly depressed at centre; surface felty or scurfy; when moist bright red-brown with striate margin, drying out to pale buff and opaque and not striate.
Stem: 4-5 cm high, 5 mm wide, tough, streaky fibrillose, reddish-brown, often twisted.
Gills: deep, thick, distant, waxy, pinkish flesh-colour with waxy appearance.
Spore-print: white.
Habitat: deciduous woodland and coniferous woodland, heaths. Often gregarious. Autumnal. Very common.

Because of its variability in appearance due to the degree of moistness of the cap, it can be very difficult to recognize in all its guises, hence the common name. The thick, distant, waxy, flesh-coloured adnate or adnexed gills are a good guide to recognition. *L. proxima* is a more robust species of damp boggy situations with stem up to 8 cm high.

COLLYBIA

Approximately 36 species, but the small dingy, grey-brown species with greyish gills are now often separated into the genus *Tephrocybe*.

COLLYBIA MACULATA
Foxy Spot
Cap: 5-9 cm diam., convex, white with pinhead-sized or larger red-brown spots, often becoming entirely pale red-brown with age.
Stem: 8-10 cm high, 1-1·6 cm wide, firm, white, often longitudinally striate, tapering below into a short rooting base.
Gills: densely crowded, shallow, pale-cream spotted with red-brown.
Habitat: coniferous woodland or amongst bracken in heathy situations where it is often gregarious and sometimes forms fairy-rings. Autumnal. Common.

The occurrence in coniferous woods or on heathland of the gregarious, tough white fruitbodies with foxy-red spotting of all parts and very crowded gills makes for easy identification.

COLLYBIA FUSIPES
Spindle Shank
Cap: 3-7 cm diam., broadly bell-shaped with central boss, smooth, dark red-brown drying pinkish buff or pale tan.
Stem: 8-10 cm high, 1-1·5 cm wide, tough, similarly coloured to cap, gradually enlarged below then conspicuously narrowed into a rooting portion, surface distinctly grooved.
Gills: adnexed, broad, distant, whitish then flushed with red-brown, also often spotted with brown.
Spore-print: white.
Habitat: tufted at base of tree trunks, especially oak. Late summer to early autumn. Occasional.

The rooting bases of the fruitbodies fuse together below ground and can be traced back to a black woody branched structure. The dark liver-coloured cap and tough grooved rooting stem are the distinctive characters.

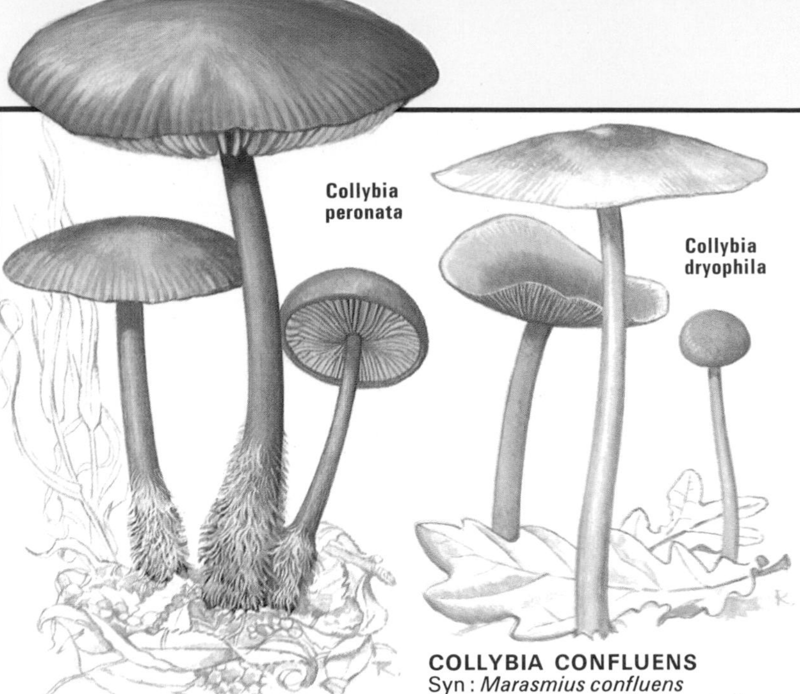

COLLYBIA PERONATA
Wood Woolly Foot
Syn : *Marasmius peronatus*
Cap: 4-6 cm diam., very broadly bell-shaped to flat, sometimes with central boss, smooth, very tough and leathery, ochre-coloured to reddish-brown, drying paler.
Stem: 7-9 cm high, 5 mm wide, tall, narrow, but very tough, pale yellowish-buff, thickly covered toward base with pale yellowish woolliness.
Gills: tough, leathery, distant, adnexed, separating from around the top of the stem in a false collar; same colour as cap.
Flesh: thin, leathery, yellowish, taste peppery.
Spore-print: white.
Habitat: especially deciduous woodland amongst leaf litter. Autumnal. Common.

The very thin leathery texture of the fruitbody makes it impossible to break; it has to be torn apart.

COLLYBIA CONFLUENS
Syn : *Marasmius confluens*
Cap: 2-4 cm diam., shallowly bell-shaped to flat, very thin, leathery, surface hoary, pale biscuit colour sometimes with hint of flesh-colour, almost white when dry.
Stem: 6-7 cm high, 3 mm wide, tall, thin, often flattened, very tough, entirely covered with minute whitish hairs, otherwise same colour as cap.
Gills: adnexed, very densely crowded, shallow, tinged colour of cap.
Spore-print: white.
Habitat: forming dense tufts of up to 10 or 15 fruitbodies in deep leaf litter in deciduous woodland, especially beech. Autumnal. Occasional.

COLLYBIA DRYOPHILA
Syn : *Marasmius dryophilus*
Cap: 2-3 cm diam., shallowly convex to flat, smooth, pale biscuit-colour, deeper brown at centre, drying out to almost white with hint of yellowish brown at centre only.
Stem: 4-6 cm high, 2-4 mm wide, tall, narrow, tough, smooth, varying from yellowish- to orange-brown.
Gills: adnexed, crowded, whitish.
Spore-print: white.
Habitat: deciduous woodland or in grassy rides. May to November. Very common.

MARASMIUS

About 30 species, mostly small to medium-sized and tough, with a white spore-print.

MARASMIUS OREADES
Fairy-Ring Champignon
Cap: 3-5 cm diam., convex with broad central boss, tough, smooth, pinkish-tan drying to pale buff, margin often grooved.
Stem: 4·5-5·5 cm high, 2·5-3 mm wide, firm, tough, pale buff.
Gills: adnexed, deep, distant, whitish.
Spore-print: white.
Habitat: pastures, lawns, roadside verges and the commonest cause of 'fairy-rings' in turf. Late summer to autumn. Common. Edible.

Recognized when forming fairy-rings by the tough, broadly bell-shaped, buff-coloured cap with distant gills. However it does not always grow in rings.

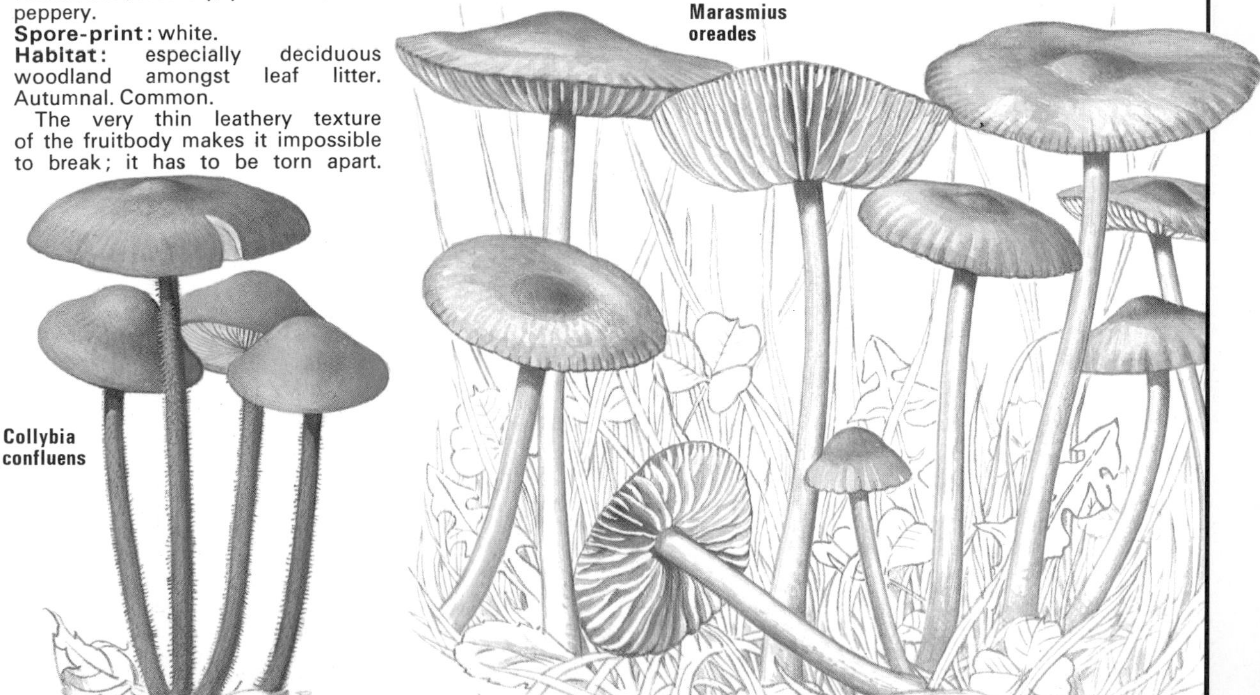

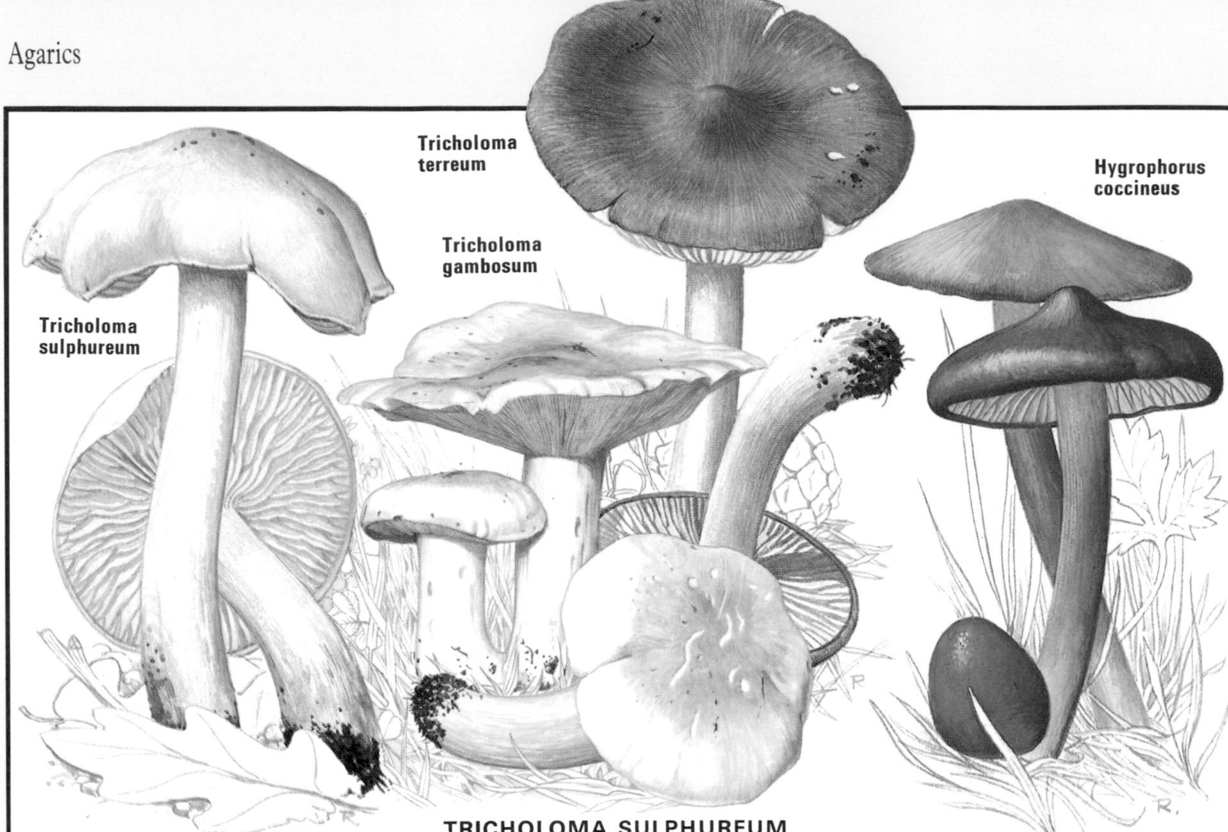

TRICHOLOMA

About 55 terrestrial species, mostly rather fleshy with sinuate gills and a white spore-print. The caps may be sticky or dry, smooth, felty or scaly and can vary in colour from bright yellow, yellowish-green, pink, violet or brown, to grey. Gills are likewise variable in colour, making it essential to check the colour of a spore-print. The stem may be smooth, fibrillose or scaly. In *T. cingulatum*, together with a very few other rare species, there is a distinct ring.

TRICHOLOMA GAMBOSUM
St George's Mushroom

Cap: 5-10 cm diam., shallowly convex, often with undulating margin, smooth, white to very pale buff at centre.
Stem: 4-7 cm high, 1·5-2 cm wide, short, squat, white.
Gills: sinuate, crowded, white.
Flesh: thick, smelling strongly of meal.
Spore-print: white.
Habitat: in pastures, roadside verges, hedge-bottoms. Appears in spring, usually around St George's Day (April 23rd), hence the common name. Edible.

This whitish mushroom is readily recognized by its squat, fleshy fruitbodies which occur in spring. *T. album* and *T. lascivum* are whitish autumnal species.

TRICHOLOMA SULPHUREUM
Gas Tar Fungus

Cap: 4-6 cm diam., convex to broadly bell-shaped, smooth, sulphur-yellow.
Stem: 7-8 cm high, 1-1·5 cm wide, same colour as cap.
Gills: sinuate, deep, distant, rather thick, same colour as cap.
Flesh: yellow with pungent smell of gas tar.
Spore-print: white.
Habitat: deciduous woodland. Autumnal. Occasional.

Easily identified by the smell and uniform sulphur-yellow colouring of the entire fruitbody.

TRICHOLOMA TERREUM

Cap: 3-6 cm, broadly bell-shaped, sometimes slightly depressed at centre, with a rather prominent nipple, mouse-grey with a felty surface.
Stem: 4-7 cm high, 7-10 cm wide, whitish.
Gills: sinuate, distant, whitish or flushed grey.
Flesh: whitish or watery-grey, mild tasting, without distinctive smell.
Spore-print: white.
Habitat: woodland, especially with conifers. Autumnal. Fairly common. Edible.

The felty grey cap and pale, distant, sinuate gills are distinctive characters.

T. argyraceum is superficially similar but smells of meal when crushed and in old specimens the gills turn yellow at the onset of decay. *T. cingulatum*, also similar and with mealy smell, is instantly recognized by having a cottony ring on the stem – one of the very few members of the genus to have one.

HYGROPHORUS

A large genus of about 90 species, many of which occur in grassland and are often of vivid colour – red, orange, yellow, pink, green, and have either a dry or glutinous surface. The cap is either shallowly convex, broadly bell-shaped, top-shaped or acutely conical. In the latter instance it often bruises black on handling or in age. Gills are strongly decurrent in many species but in others narrowly adnexed or free, but they have a waxy look. The majority of species are found in open situations, growing in short turf, but some grow in woodland.

HYGROPHORUS COCCINEUS

Cap: 2·5-5 cm diam., broadly bell-shaped, sometimes flat or even concave with central boss, smooth, bright scarlet fading with age to yellowish.
Stem: 2·5-6 cm high, 5-10 mm wide, same colour as cap but yellow at base.
Gills: adnate with decurrent tooth, rather distant, red.
Spore-print: white.
Habitat: in grassland. Autumnal. Occasional.

H. puniceus is a much more robust species with cap up to 10 cm diam., and has a white base to the stem. *H. miniatus* has a red cap with minutely scurfy surface.

HYGROPHORUS CONICUS
Cap: up to 3 cm high, acutely conical with fibrillose surface, yellow or orange becoming black on handling or with age.
Stem: up to 6 cm high, 5 mm wide, fibrillose, yellow, then blackening.
Gills: ascending, almost free, pale yellow.
Spore-print: white.
Habitat: in grassland. Autumnal. Fairly common.
The acutely conical fibrillose orange cap blackens when handled as does the rest of the fruitbody.
The only species which is liable to confusion with this species is *H. nigrescens.* This has a mainly scarlet-red cap and is more robust than *H. conicoides. H. conicus* occurs on sand dunes and has a cherry-red cap.

HYGROPHORUS PUNICEUS
Cap: 5-10 cm diam., broadly bell-shaped, smooth, scarlet but soon fading and becoming more yellowish with age.
Stem: 7-10 cm high, 1-1·5 cm wide, fibrillose, same colour as cap, but white at base.
Gills: adnexed, yellow flushed red.
Spore-print: white.
Habitat: grassland. Autumnal. Occasional.
Differs from *H. coccineus* in having a more robust appearance. *H. splendidissimus* is a similar species, but is said to have a smooth stem and also differs in the colour of the flesh in the centre of the stem: it is yellow (not whitish).

HYGROPHORUS NIVEUS
Cap: 1·5-2·5 cm diam., broadly bell-shaped to top-shaped, white and watery with striate margin when moist, opaque when dry.
Stem: 2-3 cm high, 4-5 mm wide, tapering downward; white.
Gills: decurrent, distant; white.
Habitat: in grassland. Autumnal. Common.
H. virgineus is similar but usually more robust. *H. russo-coriaceus* has a more creamy tint and a strong smell of Russian leather or of pencil sharpenings.

HYGROPHORUS PRATENSIS
Cap: 4-6 cm diam., broadly bell-shaped to top-shaped with a central boss, smooth, pale-tan to buff.
Stem: 5-6 cm high, 6-8 mm wide, same colour as cap.
Gills: decurrent, distant, pale buff.
Spore-print: white.
Habitat: in grassland. Autumnal. Occasional. Edible.
Distinguished by buff colours and distant decurrent gills.

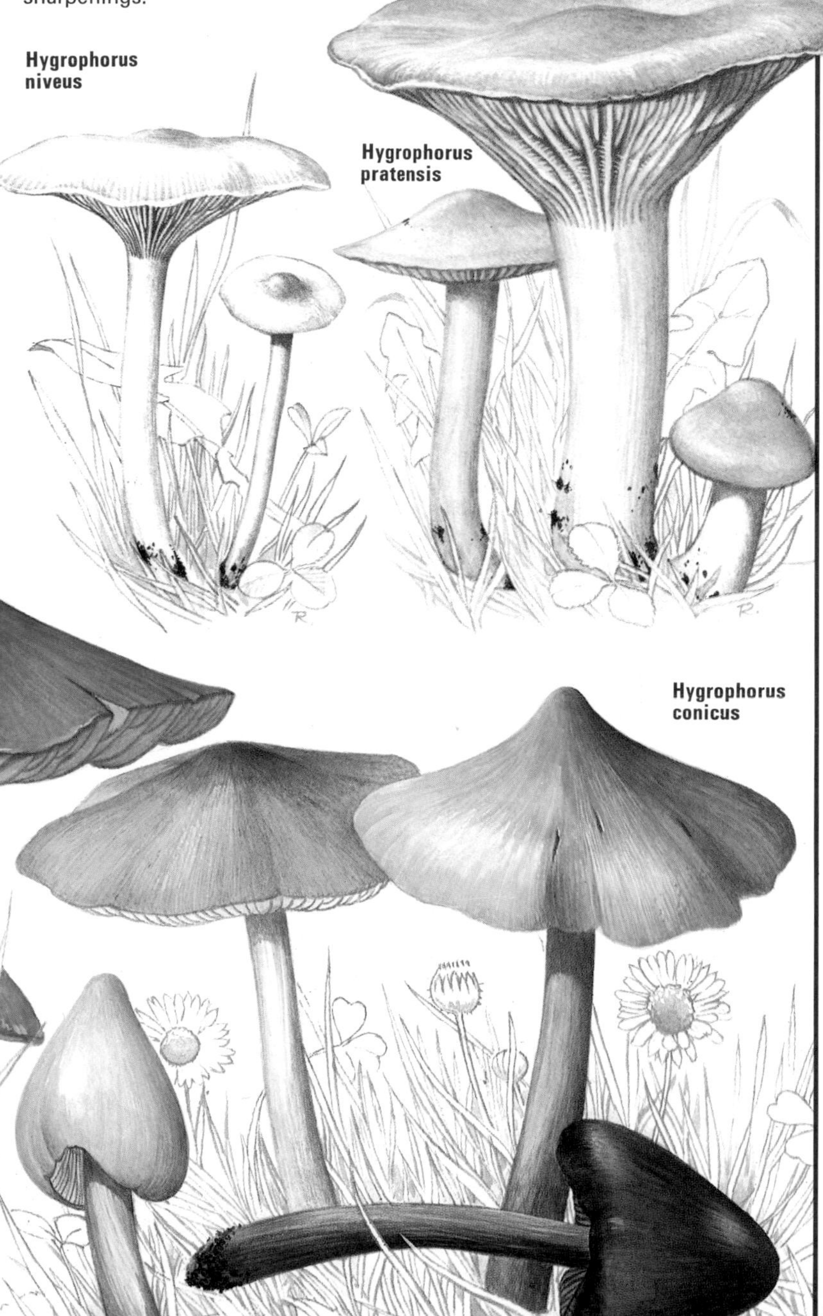

MYCENA

A large genus of over 100 species, mostly small with delicate conical to bell-shaped caps borne on a fragile elongated stem. Some species are more robust, with caps up to 4 cm in diameter and rather leathery (*M. galericulata*). A few species exude a white, red or orange latex when the stem is broken. Gills uniformly coloured or with a dark dotted gill-edge.

MYCENA INCLINATA

Cap: 1·5-2·5 cm diam., broadly bell-shaped, fairly tough, dull date brown or grey-brown, striate at margin which over-reaches the gills and is slightly toothed.
Stem: up to 7 cm high, 3 mm wide, at first silvery blue-grey and striate but fading, becoming dark tawny brown from base upward.
Gills: ascending, adnate, whitish, then flesh-colour.
Smell: distinctive, rancid.
Spore-print: white.
Habitat: densely tufted on old stumps, almost exclusively on oak, but reported on sweet chestnut and there is one authentic record on plum.

The tufted habit on oak stumps, the small, dull brown caps borne on stems with a tawny-brown base are reliable features for identification.

MYCENA GALERICULATA
The Leathery Mycena

Cap: 2-4·5 cm diam., broadly bell-shaped, flat with a central boss, tough, leathery, varying from grey-brown to buff, margin striate to somewhat grooved.
Stem: 7-10 cm high, 3-5 mm wide, tough, cartilaginous, smooth, polished, grey-brown.
Gills: adnate with decurrent tooth, deep, distant, interveined, whitish then flesh-coloured.
Smell: mealy when crushed.
Spore-print: white.
Habitat: clustered on stumps, or from buried wood. Typically autumnal, but occurring sporadically throughout the year. Very common.

Mycena inclinata differs in cap colour and tawny base to stem. *M. polygramma* has a dark smoky-brown cap which is broadly bell-shaped, 1·5 cm diam., with a central nipple and grooved margin, and firm elongated narrow stem up to 10 cm high which is silvery blue-grey and conspicuously striate. It grows singly on stumps and twigs.

MYCENA EPIPTERYGIA

Cap: 1·5-2·5 cm diam., broadly bell-shaped, sticky when moist, pale fawn with darker brown centre and radiating striae, paler when dry sometimes almost whitish.
Stem: 6-10 cm high, 2-3 mm wide, glutinous when moist, bright yellow but this often localized to base and apex, remainder pale.
Gills: adnate with decurrent tooth, white with a gelatinous edge which is separable as an elastic thread with a needle.
Spore-print: white.
Habitat: tufted in grass in woodland and on heaths. Autumnal. Fairly common.

M. viscosa is similar but occurs on conifer stumps, has a yellowish-brown cap, a more brilliant and uniformly yellow glutinous stem, a rancid smell and a tendency to become spotted with red-brown. *M. epipterygioides* also occurs with conifers but has cap and stem with green tints. *M. rorida* is another species with a glutinous stem. The cap, up to 12 mm diam., is whitish to buff-coloured with a mat, scurfy surface. The gills are white, the stem whitish and covered with a very thick transparent gluten.

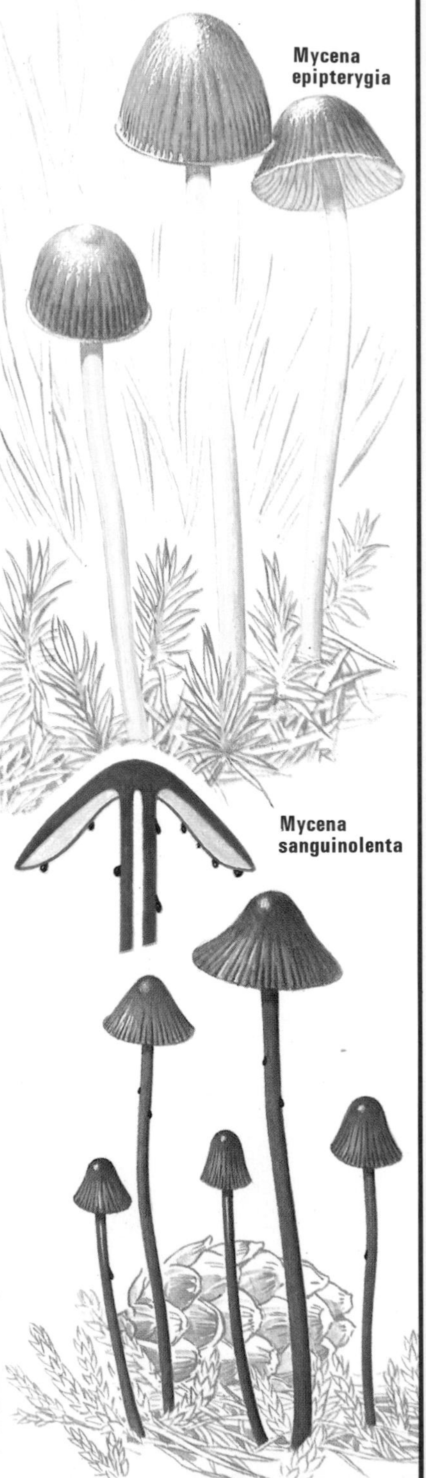

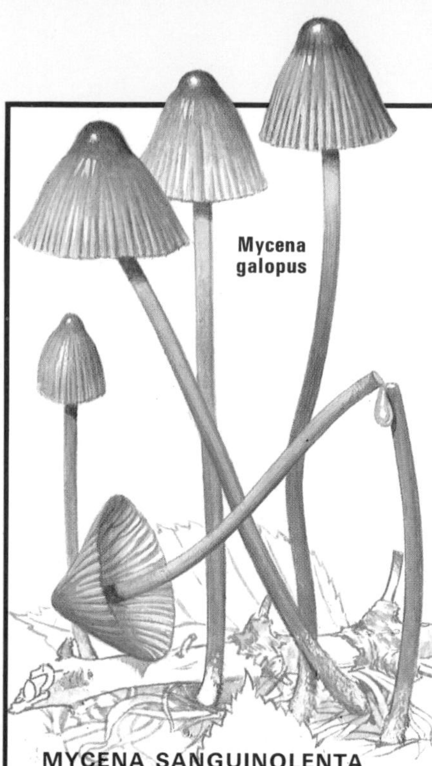

MYCENA SANGUINOLENTA
Cap: 7-10 mm high, narrowly conical, dark red-brown with striate margin.
Stem: 3-5 cm high, 1-1·5 mm wide, delicate, exuding a red juice when cut.
Gills: ascending, adnate, flesh colour with dark reddish edge.
Spore-print: white.
Habitat: in grassy places, amongst bracken, often in vicinity of pines. Autumnal. Common.

The small red-brown cap, red-dotted gill edge and red juice in the stem are distinctive. *M haematopus* also contains a red juice, but produces clustered fruitbodies on wood; the caps are dull brown with a winc rcd tint and hoary bloom. The stems are pale above and dark red-brown below. *M. crocata*, attached to twigs amongst beech leaves, exudes a bright orange juice.

MYCENA GALOPUS
Milking Mycena
Cap: 1 cm diam., broadly bell-shaped, pale with brown centre and radiating striae.
Stem: up to 5 cm high, 2 mm wide, greyish below, almost white above, when broken exuding a white milk.
Gills: adnexed, white.
Spore-print: white.
Habitat: woodland, hedge-bottoms, heaths etc. Autumnal. Very common.

The pale cap with brown radiating striae and presence of a white milk in the stem are distinctive characters. A pure white variant var. *candida* is fairly common. *M. leucogala* is distinguished by its more conical very dark brown to almost black cap and stem. *M. Galopus* is the only *Mycena* species that possesses a white milk. Note that it is best to use a young specimen to test for this milk.

RHODOTUS

A single species. Fruitbody lignicolous, fleshy, pink to apricot throughout. Cap with gelatinous wrinkled surface. Stem central, tough. Gills adnate or adnexed. Spore-print salmon-pink.

RHODOTUS PALMATUS
Cap: 3·5-6 cm diam., convex, pink at first then apricot; surface firm-gelatinous, with raised wrinkled meshwork.
Stem: 3-5 cm high, 5-10 mm wide, central to slightly excentric, paler than cap.
Gills: adnexed or adnate, deep, distant, pinkish.
Flesh: same colour as cap.
Habitat: tufted on trunks or fallen branches of elm. Autumnal but persisting into early winter. Occasional but more common than formerly owing to abundance of dead elms.

The lignicolous habit, strikingly beautiful pink colour and wrinkled gelatinous surface make for easy identification.

PLEUROTUS

About 6 lignicolous species forming solitary or clustered, tough, fleshy, shell-shaped bracket-like fruitbodies with excentric to lateral stems, often very reduced or rudimentary. Gills decurrent. Spore-print white to lilac.

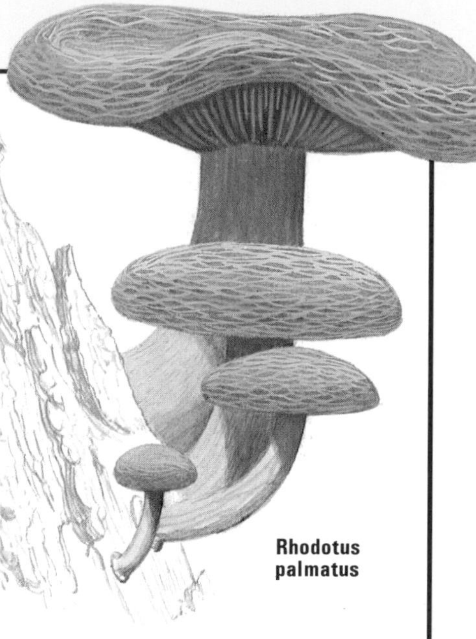

PLEUROTUS DRYINUS
Syn: *P. corticatus*
Cap: up to 15 cm wide, solitary, shell-shaped or bracket-like, whitish covered with pale-greyish felt disrupting into small rather indistinct scales especially toward the stem.
Stem: 3 cm long, 2-3 cm wide, short, lateral to very excentric, often ascending, white with an indistinct ring or ring-zone.
Gills: decurrent, distant, white.
Habitat: on live trunks of deciduous trees. Autumnal. Occasional.

The large, shell-shaped fruitbody with pale-grey scales and a ring-zone on the stem are the diagnostic features.

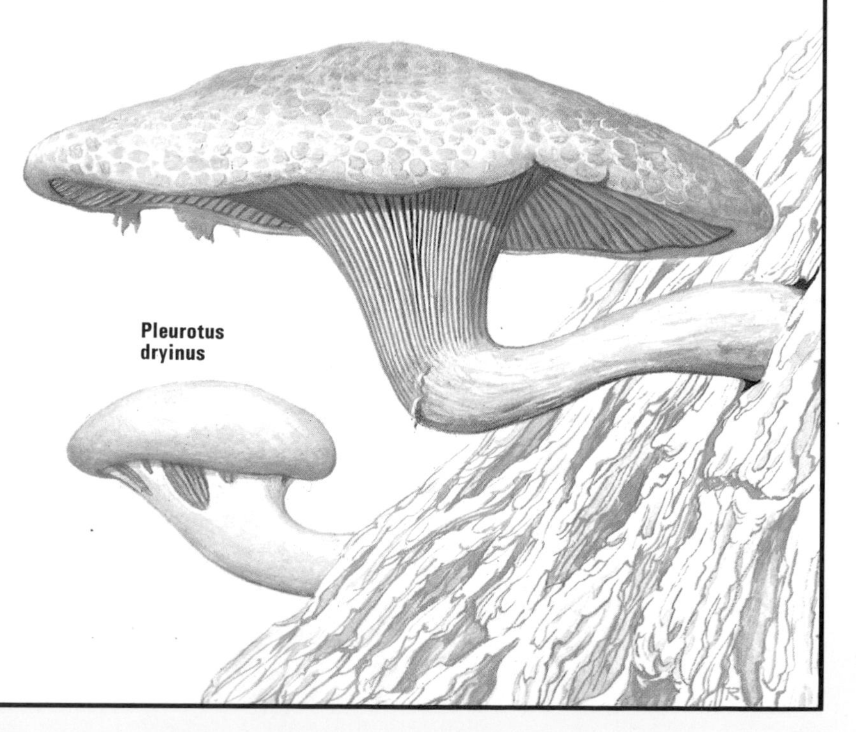

PLEURROTUS OSTREATUS
Oyster Fungus
Cap: 5-12 cm wide, tiered, convex, bracket-shaped, smooth; variable in colour from blue-grey to buff.
Stem: 2-3 cm long, 1·5-2 cm wide, short, thick, lateral, hairy, white.
Gills: decurrent, distant, whitish.
Spore-print: lilac.
Habitat: in tiered clusters on standing or fallen trunks especially beech. Autumnal but found occasionally at other times of year. Common. Edible.

The blue-grey form, which is the one to which the common name most aptly refers, is easily recognized. *P. cornucopiae* (*P. sapidus*) differs in the pale-brown to buff coloured, more or less circular, caps with depressed centre. It is often found arising from cut surfaces of stumps and roots.

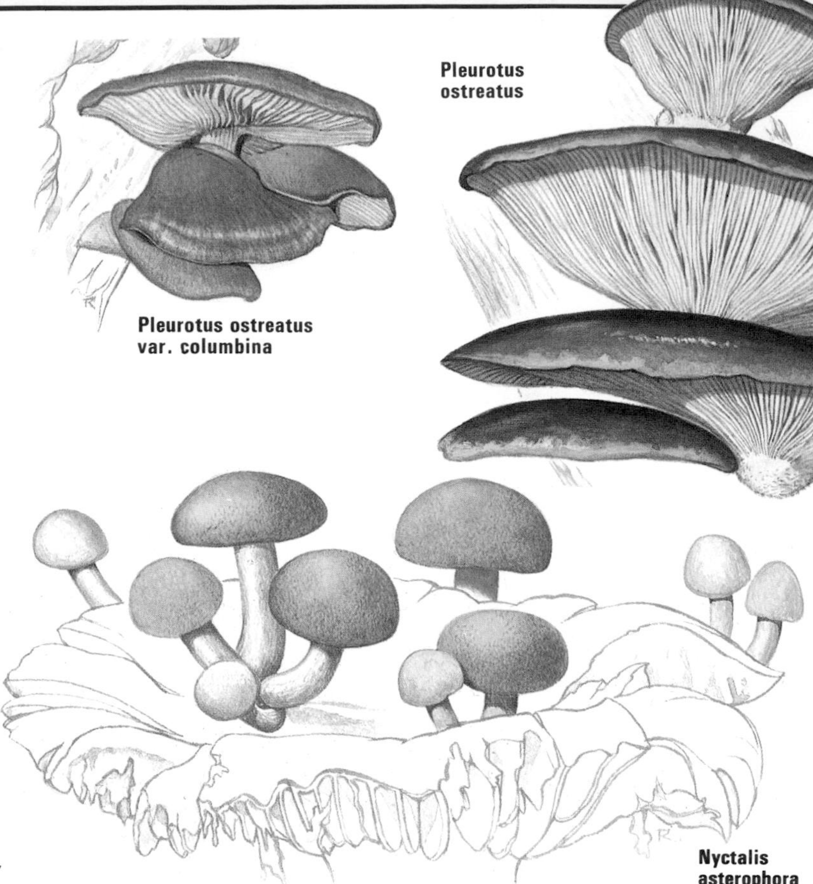

Pleurotus ostreatus var. columbina

Pleurotus ostreatus

Nyctalis asterophora

NYCTALIS
Pick-a-back Fungi

Two species growing parasitically on certain Russulas. Fruitbodies small and hemispherical with greyish cap or with cap surface brown and powdery.

NYCTALIS ASTEROPHORA
Syn: *Asterophora lycoperdoides*
Cap: 1·5-2 cm diam., hemispherical, surface brown, powdery.
Stem: 1-1·5 cm high, 2 mm wide, white.
Gills: lacking.
Habitat: clustered on rotting specimens of *Russula nigricans.* Autumnal. Rare.

Distinguished from *N. parasitica* by the powdery cap and lack of gills.

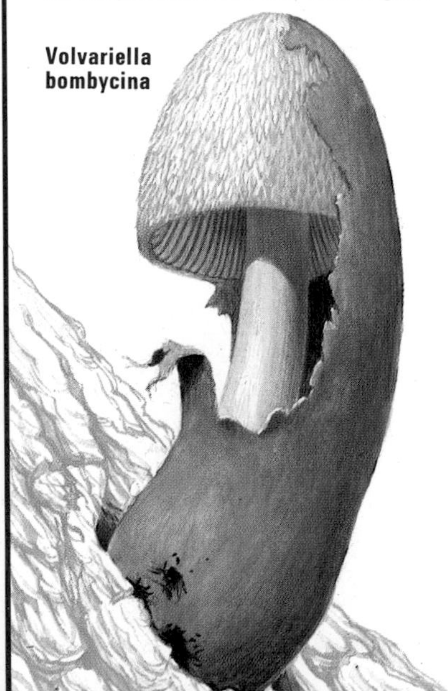

Volvariella bombycina

NYCTALIS PARASITICA
Syn: *Asterophora parasitica*
Cap: 1-1·5 cm diam., bell-shaped, flat or concave, lilac-grey.
Stem: 1·5-3 cm high, 1-2 mm wide, white.
Gills: adnate with decurrent tooth, thick, distant, often distorted, white becoming brown due to production of chlamydospores.
Habitat: clustered on various rotting *Russulas.* Autumnal. Rare.

N. asterophora differs in having a brown powdery surface to the cap and complete lack of gills.

VOLVARIELLA

Ten species ranging in size from small to robust. Cap fleshy, either dry and silky fibrillose to smooth and sticky, usually white, sometimes grey or flushed yellowish. Stem lacking a ring but with a well formed volva.

VOLVARIELLA BOMBYCINA
Syn: *Volvaria bombycina*
Cap: 6-8 cm high, conical or bell-shaped, entirely and densely covered with silky, hair-like, often upturned fibrils, white to pale-yellow.
Stem: 7-10 cm high, 1 cm wide, white.
Volva: large, sac-like, sheathing, with conspicuous free limb, dark-brown and blotched on outer surface.
Gills: free, crowded, becoming pink.
Spore-print: pink.
Habitat: on wood, generally inside hollow elm trunks. Autumnal. Rare.

Nyctalis parasitica

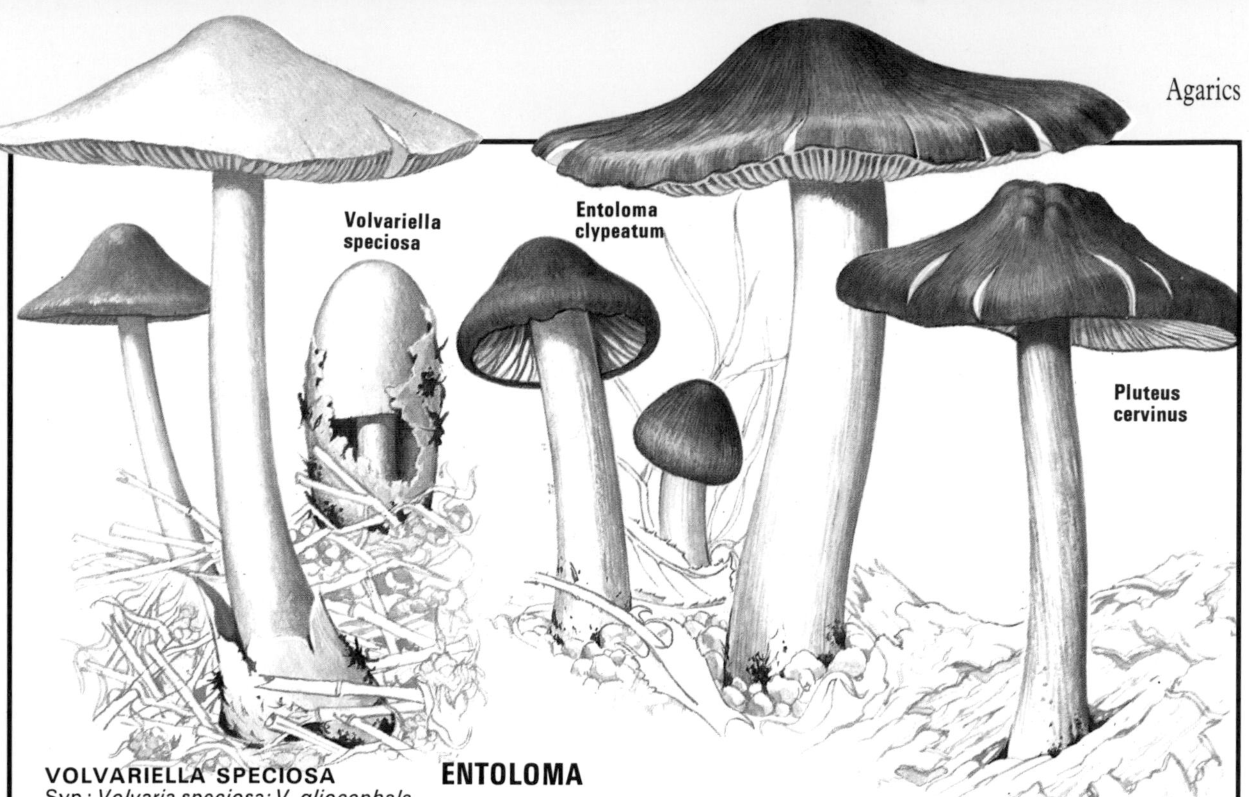

VOLVARIELLA SPECIOSA

Syn : *Volvaria speciosa; V. gliocephala*
Cap: 7-12 cm diam., at first conical, then convex with central boss, smooth, sticky, dirty-white with grey-brown centre.
Stem: 9-13 cm high, 1-1.5 cm wide, tall, narrowing above, whitish.
Volva: whitish, sac-like, but free limb relatively short and tending to collapse onto base of stem.
Gills: free, crowded, deep, white, then slowly pink with age.
Habitat: grassland, richly composted soil, near heaps of rotting grass or straw. Autumn. Occasional. Edible.

The large size, sticky pale cap with brown centre and presence of a volva are the diagnostic features.

There are several small white species of which *V. surrecta* is distinctive in growing parasitically on *Clitocybe nebularis.*

CLITOPILUS

Four species ranging from small and delicate with thin caps to fleshy, robust, fruitbodies with strongly decurrent gills. Spore-print pink.

CLITOPILUS PRUNULUS
The Miller

Cap: 3-6 cm diam., convex, then flat, top-shaped in section, smooth, dry, white to dirty-white sometimes flushed grey.
Stem: 3-5 cm high, 1-1·2 cm wide, short, thick, whitish.
Gills: decurrent, whitish at first, finally pinkish-flesh-colour.
Smell: strongly of meal, especially when bruised.
Spore-print: pink.
Habitat: deciduous woodland. Occasional. Edible.

ENTOLOMA

In the restricted sense a genus of about 25 species. Traditionally Entolomas are rather robust, fleshy, terrestrial fungi with pink sinuate gills and a pink spore-print.

ENTOLOMA CLYPEATUM

Syn : *Rhodophyllus clypeatus*
Cap: 3-6 cm diam., bell-shaped, then shallowly bell-shaped with central boss, grey-brown with darker radial streaks, drying paler.
Stem: 4-6 cm high, 6-15 mm wide, dirty-white to greyish with fibrillose surface.
Gills: sinuate, deep, distant, greyish becoming pink, edge irregularly wavy.
Flesh: greyish when water-soaked, white when dry, smelling of meal when crushed.
Spore-print: pink.
Habitat: with rosaceous shrubs and trees, hawthorn, fruit trees. Vernal.

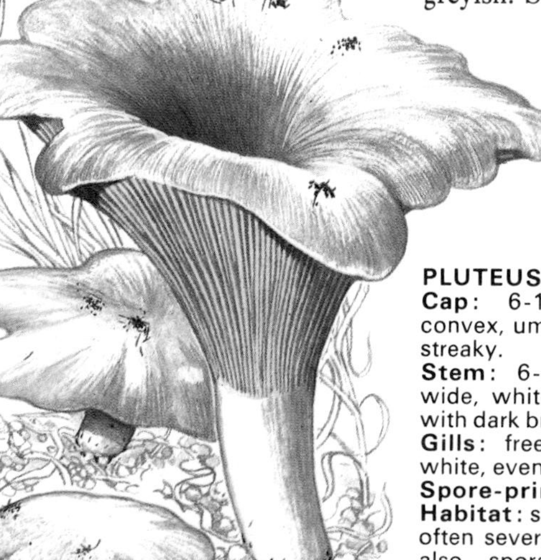

PLUTEUS

A genus of over 30 species, both lignicolous and terrestrial. Caps are smooth or fibrillose, and may appear indistinctly scaly at centre. They may also be brightly coloured in shades of yellow, orange, or greenish but are more commonly dull, brownish or greyish. Spore-print pink.

PLUTEUS CERVINUS

Cap: 6-12 cm diam., strongly convex, umber to sooty-brown, often streaky.
Stem: 6-10 cm high, 9-12 mm wide, white, lower portion speckled with dark brown fibrils.
Gills: free, crowded, rather deep, white, eventually pinkish flesh colour.
Spore-print: pink.
Habitat: stumps and sawdust heaps, often several together. Autumnal but also sporadically throughout the year. Common. Edible.

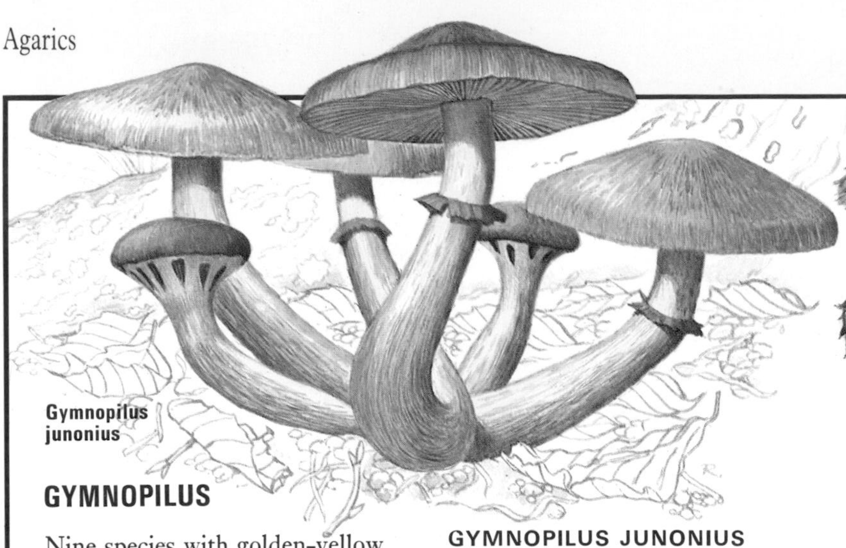

GYMNOPILUS

Nine species with golden-yellow to rich tawny caps.

GYMNOPILUS PENETRANS

Syn : *Flammula penetrans*
Cap: 3-5 cm diam., broadly bell-shaped to flattened with a central boss, smooth, golden-tawny.
Stem: 4-5·5 cm high, 3-5 mm wide, striate, fibrous, yellow above, same colour as cap below with white base, ring lacking.
Gills: adnate to slightly decurrent, yellow with rusty spots, finally tawny.
Flesh: yellow in cap, rusty-brown in stem.
Spore-print: rusty-brown.
Habitat: on twigs in coniferous woodland. Autumnal. Very common.

GYMNOPILUS JUNONIUS

Syn : *Pholiota spectabilis*
Cap: 6-12 cm diam., convex, fleshy, bright tawny or golden yellow; surface fibrillose or disrupting into indistinct fibrillose scales.
Stem: 7-15 cm high, 1·2-3 cm wide, tapering at base, fibrillose, same colour as cap but paler, and with membranous ring near apex.
Ring: yellowish, soon collapsing back onto stem.
Gills: adnate, crowded, sometimes with decurrent tooth, yellow becoming rust-coloured.
Flesh: pale yellowish.
Spore-print: rusty-brown.
Habitat: forming dense tufts at the base of trunks (sometimes living) or on stumps of deciduous trees. Autumnal. Common.

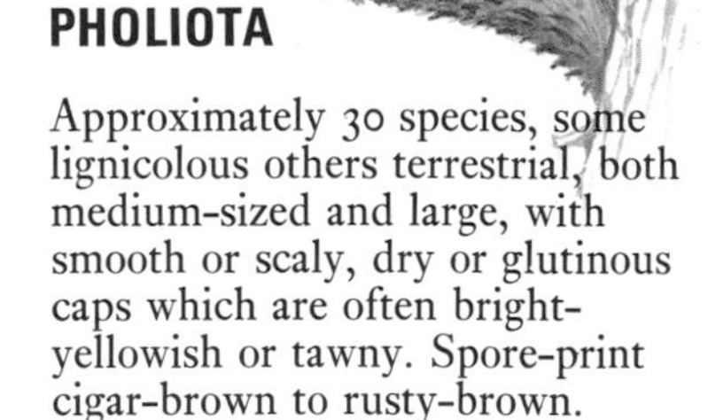

Pholiota squarrosa

PHOLIOTA

Approximately 30 species, some lignicolous others terrestrial, both medium-sized and large, with smooth or scaly, dry or glutinous caps which are often bright-yellowish or tawny. Spore-print cigar-brown to rusty-brown.

PHOLIOTA AURIVELLA

Cap: 8–15 cm diam., shallowly convex, glutinous, deep yellow with rusty-brown centre, ornamented with darker rusty-brown gelatinous scales.
Stem: 9–12 cm long, 1.5 cm wide, dry, fibrillose, yellowish becoming brownish below, with a fibrillose ring or ring-zone, below which it is often covered with pale hairy fibrils or small fibrillose scales.
Gills: adnate or sinuate, deep, pale yellow then rusty-brown.
Flesh: pale yellow, brown in base of stem.
Spore-print: rusty brown.
Habitat: in small clusters high up the trunks of deciduous trees, especially beech. Autumnal. Occasional.

PHOLIOTA SQUARROSA

The Shaggy Pholiota
Cap: 6-8 cm diam., convex, dry, pale ochre, entirely covered with prominent densely crowded up-turned bristly scales.
Stem: 6-10 cm high, 1-1·5 cm wide same colour as cap and covered with similar recurved scales below the fibrillose ring.
Gills: adnate with decurrent tooth, crowded, at first yellowish then pale rust-coloured.
Flesh: pale.
Spore-print: rusty-brown.
Habitat: parasitic on deciduous trees, forming tufts at the base of trunks. Autumnal. Fairly common.

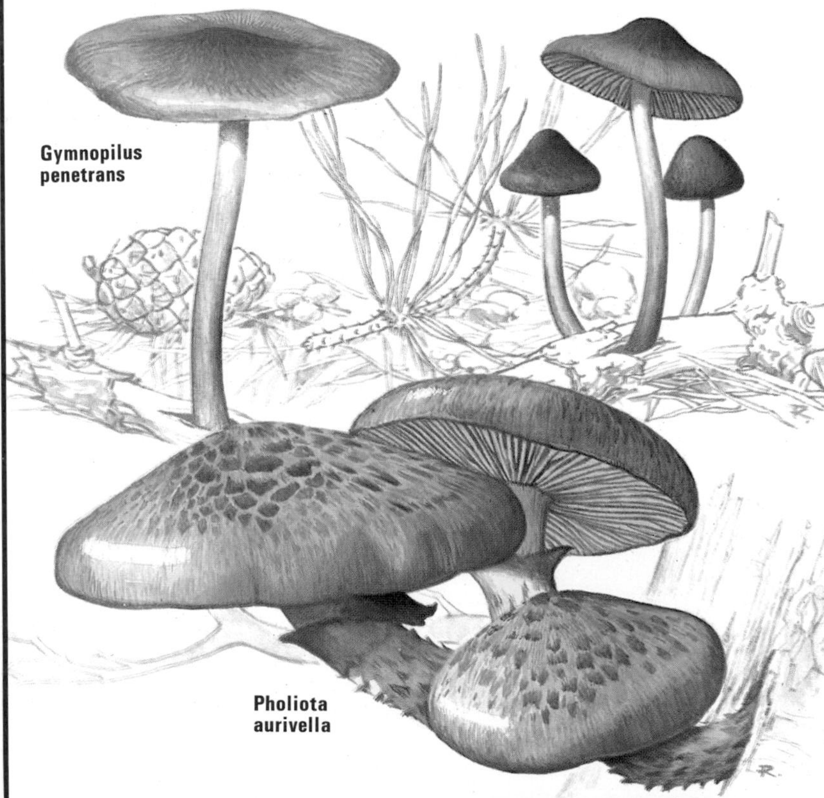

LEPTONIA

About 50 species. The genus comprises small, mostly terrestrial, species which occur mainly in short turf. The caps, seldom more than 3 cm diam., are usually convex with a depressed umbilicate centre and many have a scurfy-fibrillose or felty surface; the stems are usually long and slender.

LEPTONIA SERRULATA

Cap: 2·5 cm diam., convex, often dimpled at centre, blue-black with fibrillose surface, becoming smoky-brown with age.
Stem: 4-5 cm high, 2 mm wide, tall, slender, steely blue usually with blue-black dots at apex.
Gills: at first blue-grey, becoming pinkish but with a black-dotted edge.
Habitat: short turf. Autumnal. Occasional.

GALERINA

About 25 species, both terrestrial and lignicolous. Caps seldom more than 3·5 cm diam., convex to broadly bell-shaped, brown and conspicuously striate.

GALERINA MUTABILIS

Syn: *Pholiota mutabilis*; *Kuehneromyces mutabilis*
Cap: 2-3·5 cm diam., convex to broadly bell-shaped, watery date brown with striate margin when moist but drying out conspicuously from centre to tan colour. Fruitbodies when semi-dry are sharply two-coloured with tan centre and an abrupt, broad, watery-brown marginal zone.
Stem: 3-5·5 cm high, 3·5-5 mm wide, pale yellowish above becoming dark-brown and covered with paler recurved scales below the ring.
Gills: adnate with decurrent tooth, pale then cinnamon-brown.
Spore-print: cinnamon-brown.
Habitat: tufted on stumps of broadleaved trees. Autumnal. Fairly common. Edible.

CORTINARIUS

Our largest single genus comprising about 300 species having in common a rusty spore-print and a cobwebby cortina (ring-zone) on the stem.

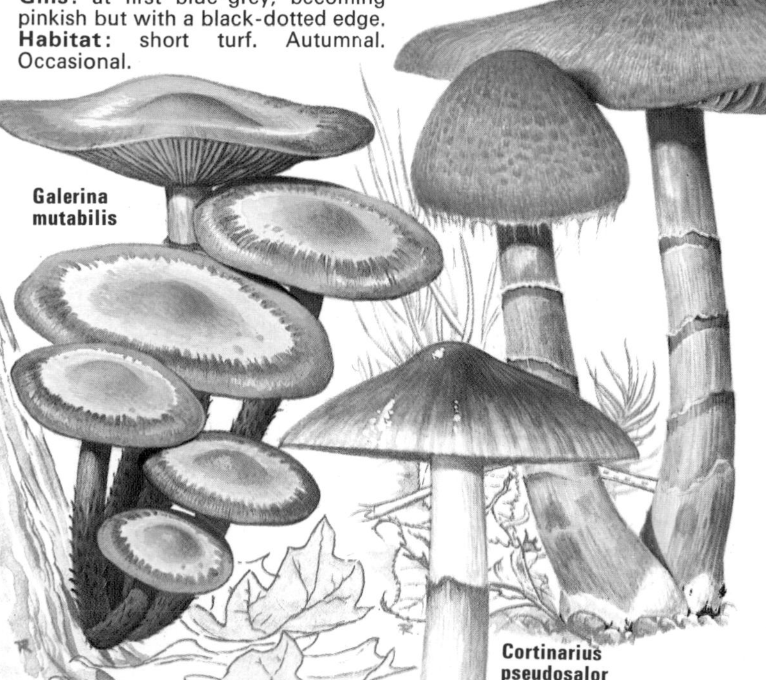

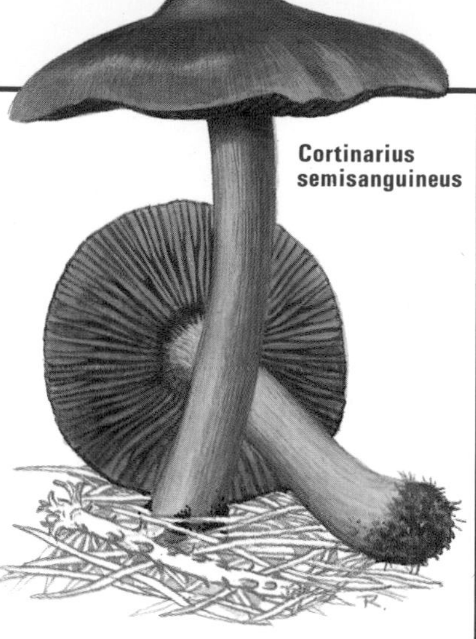

CORTINARIUS ARMILLATUS

Cap: 6-10 cm diam., shallowly convex, brick-colour to tan, smooth to indistinctly fibrillose.
Stem: 7-12 cm high, 7-12 mm wide, tall, fibrillose, with bulbous base, pale-brown with striking brick-red bands below.
Gills: adnate, cinnamon then rusty-brown.
Spore-print: rusty-brown.
Habitat: on heaths and common-land with birch trees. Autumnal. Occasional.

Distinctive on account of its size and red banding at the base of the stem.

CORTINARIUS SEMISANGUINEUS

Cap: 3-6 cm diam., shallowly convex to flat, minutely fibrillose, brown or tawny-brown tinged olive.
Stem: 4-8 cm high, 3-6 mm wide, slender, yellowish-brown often with olive tints, ring-zone yellowish.
Gills: adnate, blood red.
Flesh: yellowish brown.
Spore-print: rusty-brown.
Habitat: gregarious in coniferous woodland. Autumnal. Occasional.

Cortinarius sanguineus, which occurs in similar habitats, differs in being entirely dark blood-red.

CORTINARIUS PSEUDOSALOR

Cap: 3-5 cm diam., broadly conical to campanulate with grooved wrinkled margin, glutinous, dull violet brown with olive tint.
Stem: 5-7 cm high, 1-1·5 cm wide, tapering below, dry and white above, but covered with violet gluten from just above the mid point to the base.
Gills: adnate, crowded, becoming rusty-brown but with pale violet edge.
Spore-print: rusty-brown.
Habitat: deciduous woodland, especially with beech. Autumnal. Very common.

AGROCYBE

Twelve medium to small terrestrial or lignicolous species with smooth, convex to flat, creamy-ochre, tan to dull brown caps. Stem fairly stout to slender, with or without a ring. Gills adnate to adnexed, clay brown. Spore-print dull brown.

AGROCYBE PRAECOX

Syn: *Pholiota praecox*
Cap: 3-5 cm diam., convex, smooth, cream-coloured with ochre-coloured flush at centre.
Stem: 5-7 cm high, 5-7 mm wide, tall, slender, white, with ring.
Ring: membranous, whitish.
Gills: adnate, clay-brown, crowded.
Flesh: white in cap, brown in stem, smelling of meal when crushed.
Spore-print: clay brown.
Habitat: in grassy places, roadside verges. Vernal. Common.

Agrocybe praecox

TUBARIA

Five species with flattened cap and broadly adnate to subdecurrent brown gills, and growing either on twigs or on the ground.

TUBARIA FURFURACEA

Cap: 2·5-4 cm diam., convex then flat, cinnamon-brown when moist, drying out to pale biscuit colour from centre, but margin long remaining brown, striate and water-soaked, often with a broken ring of pale tufted scales near the edge.
Stem: 3·5-5 cm, same colour as cap but paler, sometimes with rather indistinct ring-zone.
Gills: very broadly adnate to subdecurrent, cinnamon-brown.
Spore-print: yellowish-brown.
Habitat: on twigs in deciduous woodland, or in hedge-bottoms. Autumnal, but persisting into winter and occurring sporadically throughout the year. Very common.

BOLBITIUS

A single species. Fruitbody tall, delicate, with thin flat slimy fluted cap, bright yellow at centre, brown and watery at margin. Gills free or adnexed.

BOLBITIUS VITELLINUS

Cap: 2·5-3·5 cm diam., at first acorn-shaped, becoming flat with strongly grooved margin, surface smooth, sticky, chrome yellow at first, but this colour later restricted to centre with the margin brown and watery.
Stem: 6-10 cm high, 3-5 mm wide; pale yellow to whitish with hoary surface.
Gills: free to adnexed, cinnamon-brown.
Spore-print: rusty-brown.
Habitat: solitary or gregarious in richly manured grass and amongst wood chips. Autumnal. Occasional.

INOCYBE

A large genus of about 86 terrestrial, small to medium-sized species, most of which cannot be identified without a microscope.

INOCYBE ASTEROSPORA

Cap: 3-4·5 cm diam., shallowly convex with central boss, fawn-brown to chestnut at centre but toward margin surface disrupting into radial fibrils, same colour as cap, on a pale background.
Stem: 3·5-6 cm high, 5 mm wide, cylindrical with conspicuous basal bulb having a well-defined margin; surface striate, entirely pruinose, fawn-brown, paler at apex.
Gills: adnexed, becoming cinnamon to clay-brown.
Habitat: deciduous woodland. Autumnal. Occasional. POISONOUS.

Recognized by its relatively tall, brown stem with conspicuous white basal bulb and brown, radially fissured cap.

INOCYBE GEOPHYLLA

Cap: 1-2·5 cm diam., conical or bell-shaped, with white, silky fibrillose surface.
Stem: 2-3·5 cm high, 3-4 mm wide, white, pruinose at apex.
Gills: adnexed, white, slowly becoming clay-brown.
Spore-print: clay-brown.
Habitat: deciduous woodland. Autumnal. Common. POISONOUS.

INOCYBE GEOPHYLLA LILACINA

As above but entirely lilac except sometimes for a yellowish tint at centre, and for the gills ultimately becoming clay-brown.

Bolbitius vitellinus

Inocybe asterospora

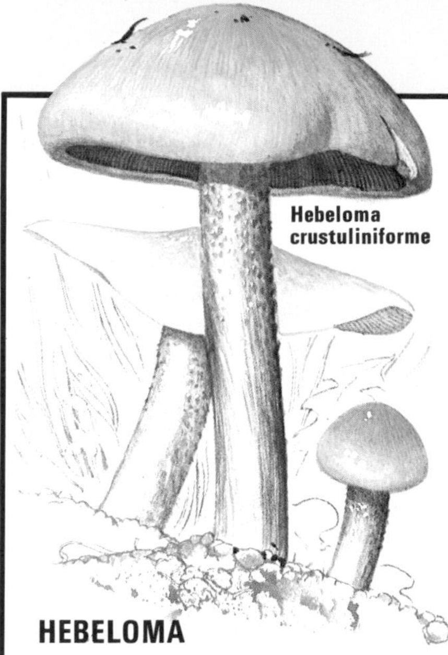

HEBELOMA

About 20 terrestrial species, mostly fleshy, varying from small to large, usually with smooth, pale biscuit-coloured caps often with a flush of brown or tan at centre and with a sticky surface.

HEBELOMA CRUSTULINIFORME
Fairy Cake Fungus

Cap: 3-5·5 cm diam., convex, pale biscuit-coloured with flush of tan at centre, surface sticky.
Stem: 3·5-5 cm high, 5-8 mm wide, white, pruinose above.
Gills: sinuate becoming clay-brown, often with watery droplets along the edge in damp weather.
Flesh: white, smelling strongly of raddish.
Spore-print: clay-brown.
Habitat: deciduous woodland. Autumnal. Fairly common. POISONOUS.

CREPIDOTUS

A genus of 18 species, growing on wood, herbaceous plant material, rarely on soil. Cap shell-shaped ranging from about 1 cm diam. to 5 cm diam., mostly white or whitish, occasionally watery brown, sometimes with a separable gelatinous skin. Spore-print pinkish-brown to snuff-brown.

The small white species of *Pleurotus* or *Pleurotellus* are at once distinguished by having a white print.

CREPIDOTUS MOLLIS

Cap: 2-5 cm diam., horizontal, shell-shaped, growing in tiers; attached directly to wood, soft, flabby, pale watery yellowish-brown with striate margin, drying paler, surface gelatinous, separable as an elastic skin.
Spore-print: snuff-brown.
Habitat: tiered on stumps. Autumn. Occasional.

Distinguished by the shell-shaped tiered brackets with horizontal brown gills and a separable gelatinous cuticle. The latter is readily demonstrated if the cap is stretched laterally, when the transparent, gelatinous layer will be seen as an elastic film between the gills. *C. calolepis* has the cap ornamented with bright rusty-brown, fibrillose scales. The many white species can only be distinguished after microscopic examination.

PAXILLUS

Four species, some terrestrial others lignicolous.

PAXILLUS INVOLUTUS
The Roll Rim

Cap: 5-11 cm diam., convex with strongly enrolled margin, becoming shallowly depressed at centre, surface glutinous in wet weather, yellowish-brown, smooth except toward the edge which is often ribbed and downy.
Stem: 6-7 cm high, 1-1·2 cm wide, central, same colour as cap but paler, often streaky.
Gills: decurrent, yellowish-brown becoming dark red-brown when bruised.
Spore-print: ochre-brown.
Habitat: heathy places, associated with birch. Autumnal. Very common. POISONOUS.

Easily recognized by its association with birch, the brown glutinous cap with enrolled, woolly, ribbed margin.

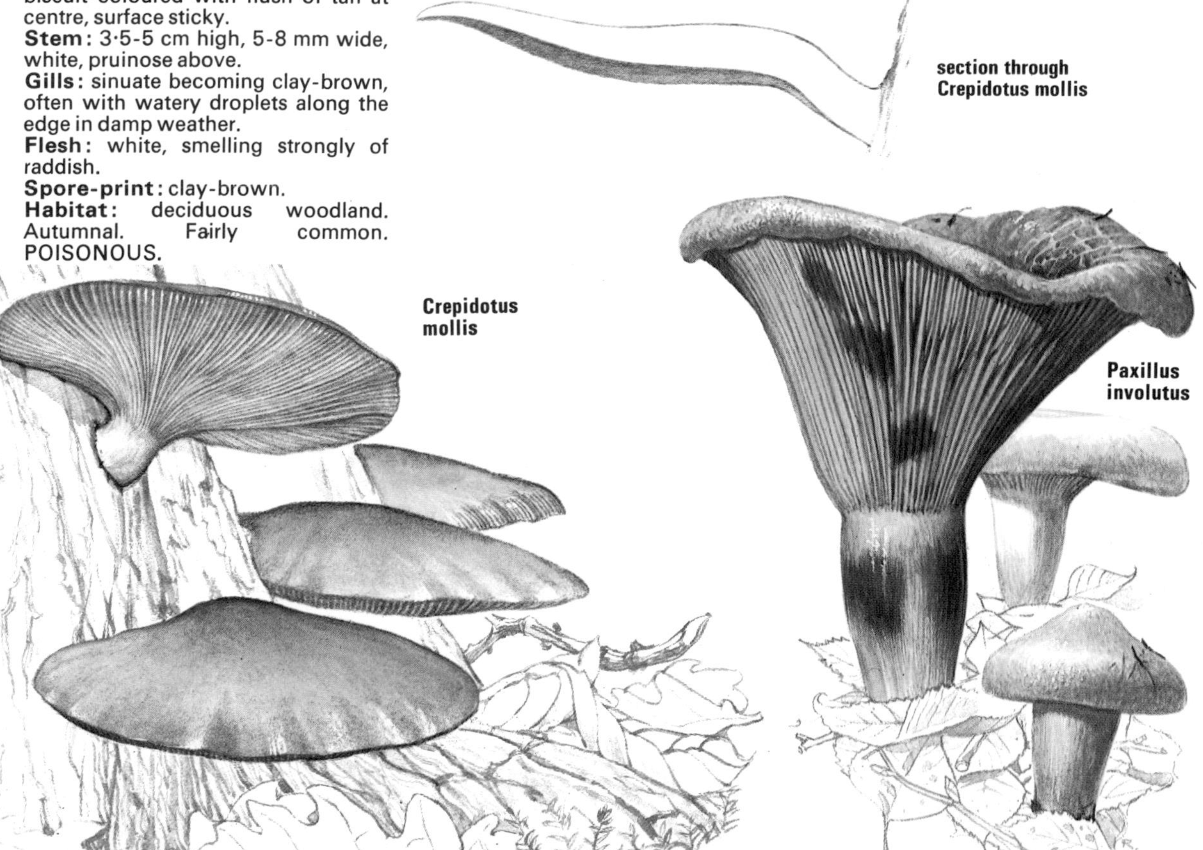

COPRINUS
Ink Caps

About 90 species, large or minute, mostly growing on dung or richly manured ground, sometimes at the base of stumps or from roots and then often tufted. Cap surface usually white or grey, more rarely brown or some other colour, either powdery like meal or with fibrillose scales. Stem with or without a ring. Gills black, usually liquefying after a few hours giving an ink-like fluid.

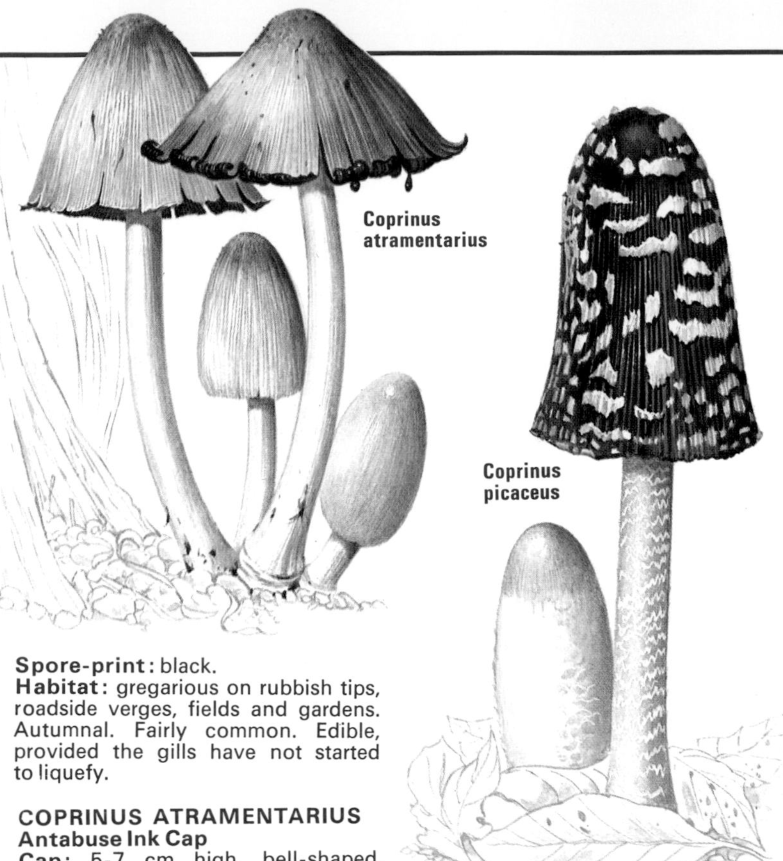

COPRINUS PLICATILIS
The Little Japanese Umbrella
Cap: 2-3 cm diam., acorn-shaped, yellowish-brown and closely striate at first, then flat, coarsely grooved or fluted, grey with small depressed tan-coloured disc, very thin, short-lived, almost translucent.
Stem: 6-8 cm high, 3 mm wide, whitish, very delicate.
Gills: free, attached to a collar around stem apex, scarcely liquefying.
Spore-print: black.
Habitat: damp grass, lawns, road-side verges. Autumnal. Common.

COPRINUS COMATUS
Shaggy Ink Cap; Lawyers Wig
Cap: 6-14 cm high, cylindrical, opening slightly at base, eventually bell-shaped, white, with buff-coloured central disc, surface broken up into shaggy scales, margin closely striate becoming greyish, finally black on maturity. Entire cap gradually dripping away from margin as an inky fluid.
Stem: up to 30 cm high, 1-1·5 cm wide, white with movable membranous ring toward base.
Gills: free, crowded, white, near stem apex, then pink and finally black at margin.
Spore-print: black.
Habitat: gregarious on rubbish tips, roadside verges, fields and gardens. Autumnal. Fairly common. Edible, provided the gills have not started to liquefy.

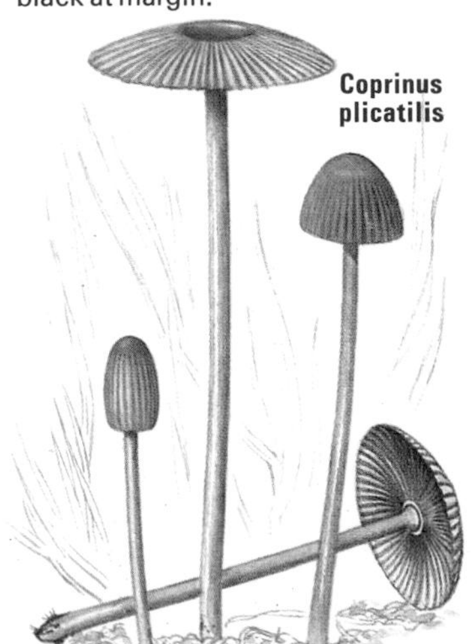

COPRINUS ATRAMENTARIUS
Antabuse Ink Cap
Cap: 5-7 cm high, bell-shaped, irregularly ribbed and wrinkled almost to disc, grey, often with few inconspicuous brownish scales at centre.
Stem: 7-9 cm high, about 1 cm wide at base, where there is a conspicuous oblique ring-zone, white.
Gills: free, whitish, becoming grey, finally black, liquefying.
Spore-print: black.
Habitat: tufted in vicinity of stumps of deciduous trees or from roots, often in gardens, fields etc. Autumnal but sporadically throughout the year from early spring. Common. Edible, but causing sickness if eaten with alcohol due to a substance contained in the fungus identical with the drug Antabuse, which is used in treatment of chronic alcoholics.

This robust, tufted, greyish ink cap is readily identified. *C. acuminatus* also tufted or solitary, is a woodland fungus which is less robust and has a darker, more egg-shaped cap.

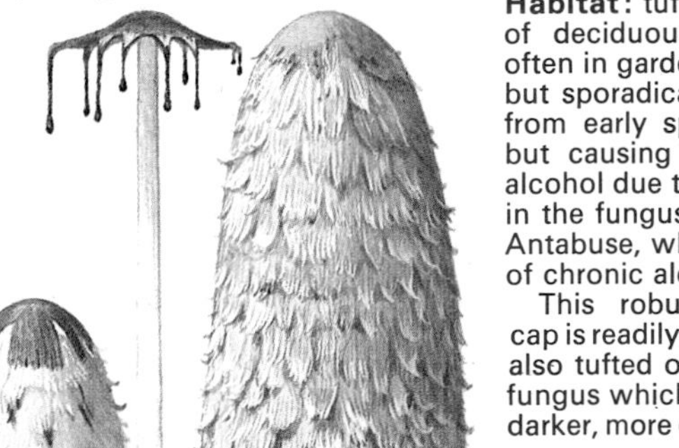

COPRINUS PICACEUS
The Magpie
Cap: 8 cm high, oval, then bell-shaped, cap surface disrupting into conspicuous white patches on a blackish background.
Stem: up to 25 cm high, 1 cm wide, hollow, fragile, white, lacking a ring.
Gills: free, crowded, white above, then pinkish, buff and finally black at margin, liquefying readily.
Spore-print: black.
Habitat: beech woods. Autumnal.

Easily identified by the robust stature and cap with contrasting white patches on a black background. This black and white colouring has led to the popular name 'The Magpie'.

COPRINUS DISSEMINATUS
The Trooping Crumble Cap
Syn : *Psathyrella disseminata*
Cap: 5-10 mm high, acorn-shaped to hemispherical, pale yellowish-clay becoming greyish at the margin, closely grooved almost to the centre, minutely hairy under a very strong lens.
Stem: 1-3·5 cm high, 1-1·5 mm wide, white, very delicate and brittle.
Gills: adnate, dark grey to blackish, scarcely liquefying.
Spore-print: black.
Habitat: densely gregarious, covering entire stumps in myriads of tiny, brittle, bell-shaped, biscuit-coloured fruitbodies. Autumnal, but also sporadically throughout the year. Common.

Coprinus disseminatus

AGARICUS

About 40 small, medium- or large-sized, terrestrial species with cap surface either white or brown. Stem easily separable from cap, and with membranous ring. Ring simple or appearing cog-wheel-like below. Gills free, finally deep purple-brown. Flesh unchanging, reddening or becoming yellow when bruised. Spore-print purple-brown or chocolate-brown. Before eating, it is essential to check that the gills are purple-brown at maturity, and that there is a ring on the stem but no volva.

AGARICUS CAMPESTRIS
Field Mushroom
Syn : *Psalliota campestris*
Cap: 4·5-8 cm diam., convex then flat, white, surface often disrupting into indistinct fibrillose scales especially around centre.
Stem: 4-6 cm high, 1-1·5 cm wide, short, squat, with pointed base and ring.
Ring: poorly developed, simple, often little more than a torn fringe.
Gills: free, at first pink then purplish-brown.
Flesh: white, sometimes reddish in stem when cut.
Spore-print: purplish-brown.
Habitat: grassy places, pastures, often in fairy-rings. Autumnal. Occasional. Edible.

Distinctive characters are the squat growth form, pink colour of the young gills which finally become purple-brown and the poorly developed ring.

AGARICUS ARVENSIS
Horse Mushroom
Syn : *Psalliota arvensis*
Cap: 6-11 cm diam., hemispherical, becoming shallowly convex, white, often creamy with age, sometimes faintly yellow when handled, but never vividly so.
Stem: 8-12 cm high, 1·5-2 cm wide, tall, cylindrical, with bulbous base, white with membranous ring.
Ring: large, pendulous, high on stem, underside like radiating spokes of a wheel (cog-wheel-like).
Gills: white, then brownish, finally purplish-brown, never pink.
Flesh: white, unchanging.
Spore-print: purplish-brown.
Habitat: open grassland, hillsides, orchards etc, often in fairy-rings. Autumnal. Occasional. Edible.

AGARICUS BITORQUIS
Syn: *Psalliota bitorquis; Agaricus edulis*
Cap: 6-10 cm diam., shallowly convex with enrolled margin, whitish to pale buff.
Stem: 4-11 cm high, 2-3 cm wide, short, stout, sheathed below to an inferior ring.
Ring: double, the upper thicker and better developed, the lower narrower, thinner sometimes reminiscent of a volva, sometimes disrupting into bands of scales.
Gills: free; dirty-pink, then purple-brown.
Flesh: white, reddening slightly when broken.
Spore-print: purple-brown.
Habitat: often in soil around the base of trees in pavements, sometimes pushing up paving stones, and roadsides. Autumnal. Occasional. Edible.

The inferior double ring is diagnostic. Check that the spore-print is purple-brown. Specimens which stain bright yellow should be avoided (see *A. xanthodermus*).

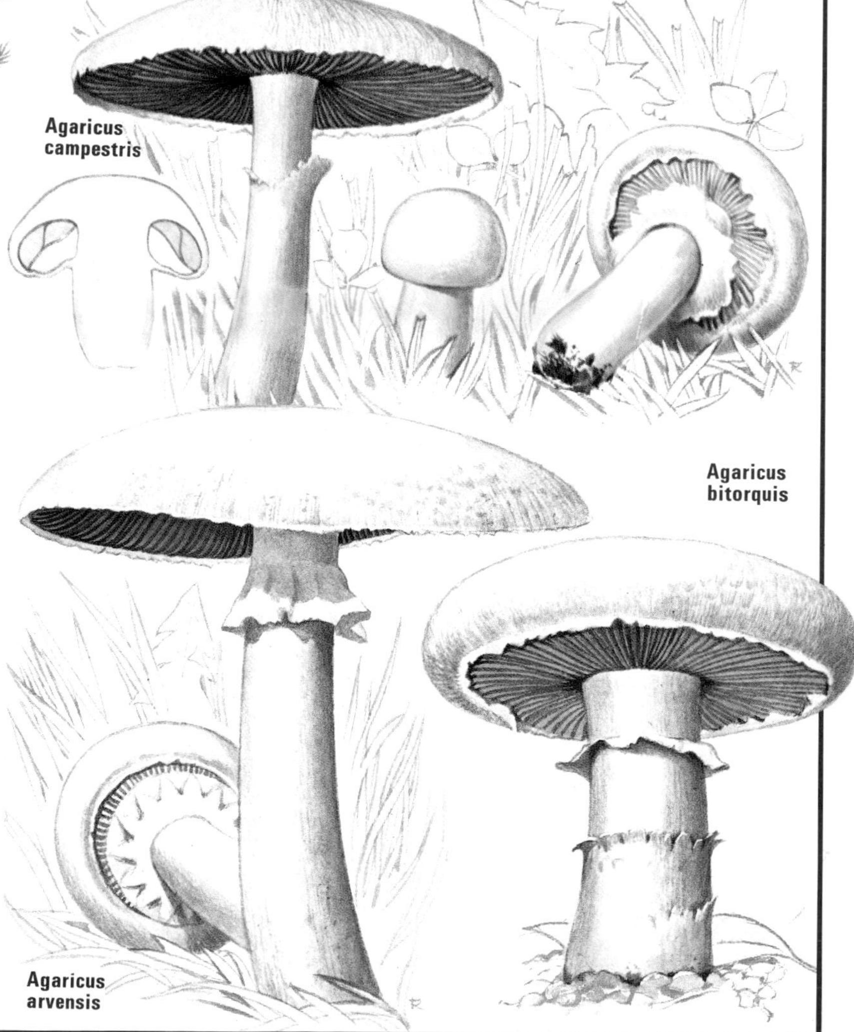

AGARICUS PLACOMYCES
Syn: *Psalliota placomyces; Agaricus meleagris*
Cap: 5-8 cm diam., shallowly convex with slightly depressed centre, surface disrupting except at disc into small, densely crowded, greyish-black scales on a whitish background; the disc is uniformly grey-black; on handling the cap bruises vivid yellow.
Stem: 7-9 cm high, 8-12 mm wide, rather tall and narrow, white, bruising yellow, with ring.
Gills: at first pink, then purplish-brown or blackish-brown.
Flesh: white, vivid, yellow in stem base.
Spore-print: purplish-brown.
Habitat: deciduous woodland. Autumnal. Occasional.

The blackish scaly cap and yellow staining which develops on bruising are distinctive characters. *Agaricus langei, A. haemorrhoidarius* and *A. silvaticus* are all species with brown scaly caps and reddening flesh.

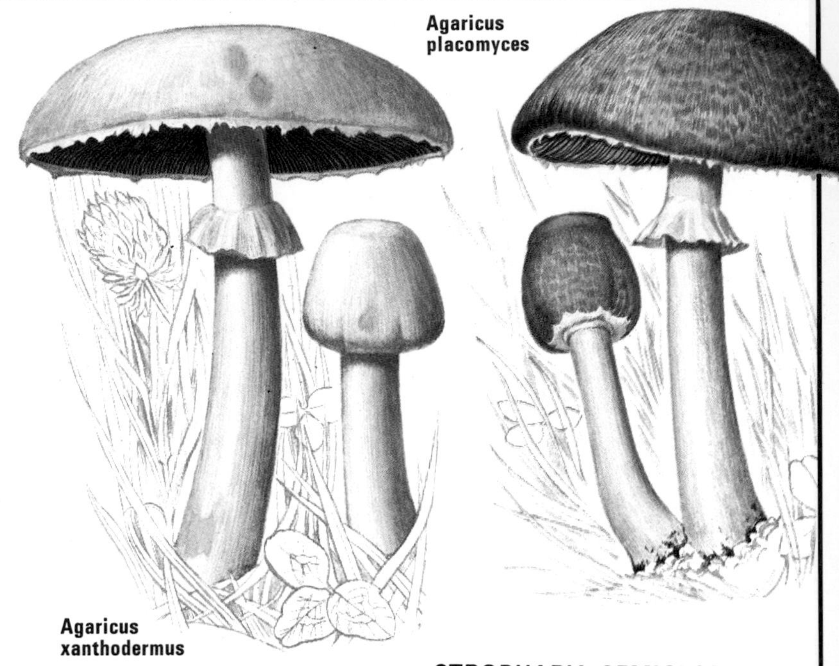
Agaricus placomyces
Agaricus xanthodermus

AGARICUS XANTHODERMUS
Yellow-Staining Mushroom
Syn: *Psalliota xanthoderma*
Cap: 5-8 cm diam., at first hemispherical with truncate top, later broadly bell-shaped to shallowly convex; white; often cracked or fissured toward margin; becoming vivid yellow when scratched.
Stem: 6-7 cm high, 1-1·5 cm wide; cylindrical with bulbous base; white, becoming yellow when scratched; ring present.
Ring: membranous.
Gills: whitish, then grey, finally purple-brown.
Flesh: white becoming vivid yellow in base of stem.
Spore-print: purple-brown.
Habitat: gregarious in pastures, woodland, gardens, shrubberies. Autumnal. Occasional. POISONOUS.

The white cap and yellow staining enable this species to be readily identified.

STROPHARIA

About 15 terrestrial species of small to medium size, often with glutinous cap. Stem with membranous ring or prominent ring-zone. Gills sinuate or adnate. Spore-print black.

Species of *Agaricus* differ in having a dry cap, free gills and a purple-brown spore-print.

STROPHARIA AERUGINOSA
Verdigris Fungus
Cap: 3·5-5 cm diam., shallowly bell-shaped, glutinous, bright blue-green with tiny white fleck-like scales floating in the gluten toward the margin; with age the scales and gluten wash off leaving cap yellowish.
Stem: 4-6 cm high, 3-5 mm wide, with ring near apex, smooth and whitish above ring, pale blue-green and cottony-scaly below.
Ring: sometimes collapsing to blackish ring-zone indicated by trapped spores.
Gills: sinuate, smoky-brown with white edge.
Spore-print: black.
Habitat: deciduous woodland, gardens, often in nettle beds. Autumnal. Occasional.

Unmistakable due to the blue-green, glutinous cap.

STROPHARIA SEMIGLOBATA
Cap: 1-2·5 cm diam., hemispherical, pale yellowish to yellowish-tan, glutinous when wet, shiny when dry.
Stem: 4-8 cm high, 1-2 mm wide, tall, narrow, fragile, with black ring-zone near apex, dry above ring, glutinous below.
Gills: adnate, deep, dark-brown.
Habitat: especially on horse-droppings, richly manured places such as gardens and pastures. Autumnal. Locally common.

S. merdaria is similar but has a more brownish, flatter cap and an entirely dry stem.

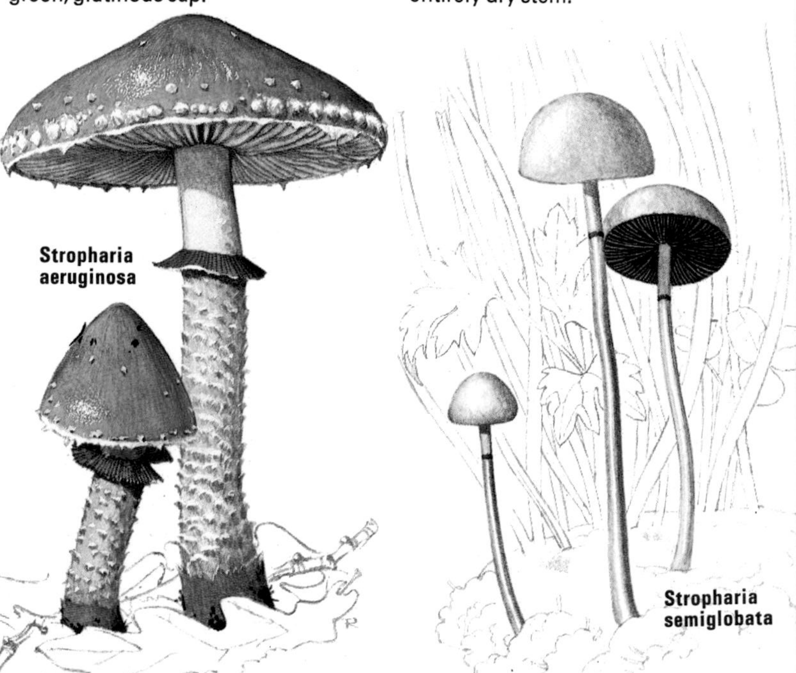
Stropharia aeruginosa
Stropharia semiglobata

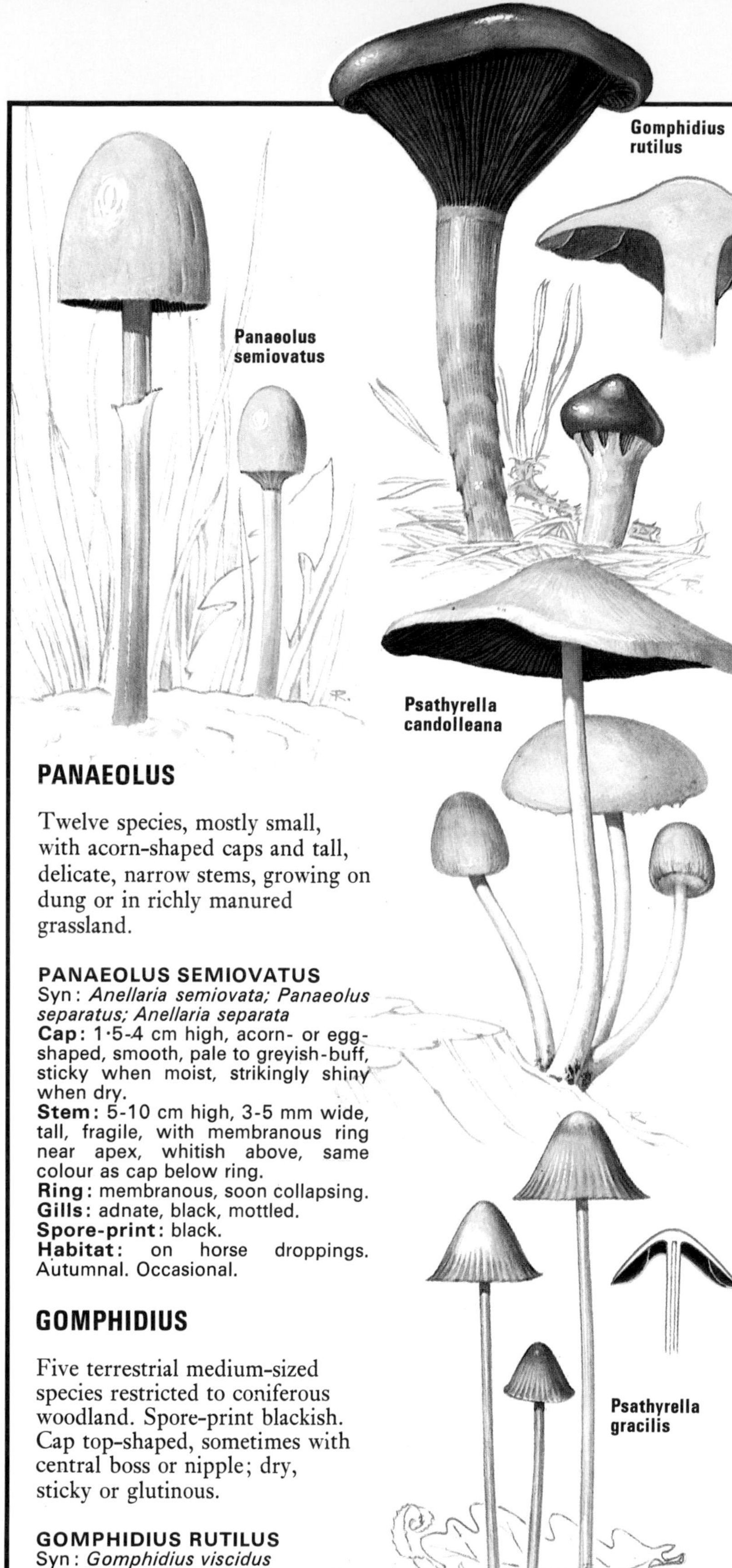

PANAEOLUS

Twelve species, mostly small, with acorn-shaped caps and tall, delicate, narrow stems, growing on dung or in richly manured grassland.

PANAEOLUS SEMIOVATUS

Syn: *Anellaria semiovata; Panaeolus separatus; Anellaria separata*
Cap: 1·5-4 cm high, acorn- or egg-shaped, smooth, pale to greyish-buff, sticky when moist, strikingly shiny when dry.
Stem: 5-10 cm high, 3-5 mm wide, tall, fragile, with membranous ring near apex, whitish above, same colour as cap below ring.
Ring: membranous, soon collapsing.
Gills: adnate, black, mottled.
Spore-print: black.
Habitat: on horse droppings. Autumnal. Occasional.

GOMPHIDIUS

Five terrestrial medium-sized species restricted to coniferous woodland. Spore-print blackish. Cap top-shaped, sometimes with central boss or nipple; dry, sticky or glutinous.

GOMPHIDIUS RUTILUS

Syn: *Gomphidius viscidus*
Cap: 4-8 cm diam., convex often with central nipple, smooth, sticky, brown with purplish or wine-red tints, margin enrolled and yellowish or coppery.
Stem: 6-10 cm high, 1-1·5 cm wide, tapering below, fibrillose, purplish brown like cap, with broad cottony, buff-coloured ring-zone on upper portion.
Gills: decurrent, distant, yellowish-grey finally dingy purplish.
Flesh: entirely yellowish-tan.
Spore-print: blackish.
Habitat: pine woods. Autumnal. Occasional.

PSATHYRELLA

A genus of over 60 species, both terrestrial and lignicolous, mostly small to medium-sized. Spore-print dark purplish-brown to almost black. Often very fragile, with caps commonly or broadly bell-shaped, sometimes flattened, either naked or with whitish or blackish scales, surface usually drying out rapidly and becoming very pale.

PSATHYRELLA CANDOLLEANA

Syn: *Hypholoma candolleanum*
Cap: 2·5-5 cm diam., shallowly bell-shaped to flat, pale creamy-ochre to whitish especially when dry, with tiny tooth-like remnants of veil hanging from margin when young.
Stem: 4-6 cm high, 4-5 mm wide, white, hollow, very brittle.
Gills: adnate, for a long time whitish then lilac-grey and finally brownish-black.
Spore-print: almost black.
Habitat: tufted on wood, stumps, roots, fence-posts. Spring to Autumn. Common.

The pale, flattened, tufted, fragile fruitbodies with glistening surface to the cap and lilac-grey gills are easy to identify. *P. hydrophila* is another densely tufted species, formerly more common than at present, with watery date-brown, striate caps, becoming opaque and pale tan on drying.

PSATHYRELLA GRACILIS

Cap: 1·5-2·5 cm diam., bell-shaped, dark brown or reddish-brown with striate margin when moist, opaque and pale biscuit colour when dry.
Stem: 8-10 cm high, 2 mm wide, very tall, fragile, pure white with white hairy fibrils at base.
Gills: adnate, blackish, with pink edge.
Habitat: solitary or gregarious, amongst grass or leaves, in deciduous woodland, hedge-bottoms, roadside verges. Autumn. Very common.

The tall, fragile white stem, small bell-shaped cap, often seen in the pale condition, with glistening surface, together with the dark gills edged with pink are the salient features.

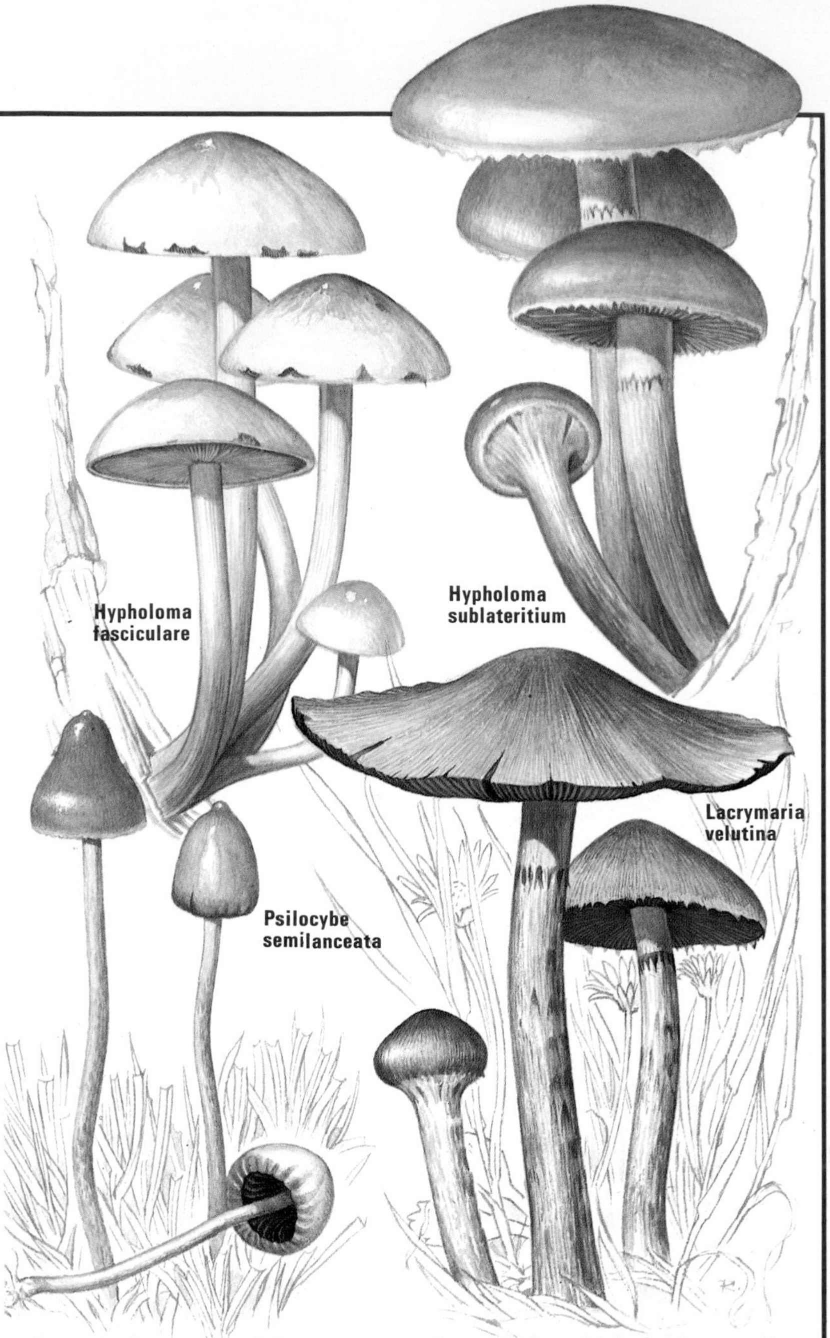

PSILOCYBE

About 20, mostly small terrestrial species of heaths and grassland, a few growing on twigs and sawdust. Gills adnexed or adnate, sometimes broadly so, reddish or purplish-brown.

PSILOCYBE SEMILANCEATA
Liberty Cap
Cap: 1-1·5 cm high, conical with sharp-pointed apex, sticky when moist, pale tan drying out to pale creamy-buff.
Stem: 3·5-5 cm high, 2 mm wide, paler than cap, sometimes blue at base.
Gills: adnate or adnexed; ascending; black with white edge.
Spore-print: purple-brown to almost black.
Habitat: often gregarious in grassland, lawns, fields, heaths. Autumnal. Fairly common. Hallucinogenic.
The characteristic shape of the tiny cap affords easy recognition.

LACRYMARIA

Two tufted species, apparently terrestrial, but growing from buried wood.

LACRYMARIA VELUTINA
Weeping Widow
Syn: *Hypholoma velutinum.*
Cap: 4-6 cm diam., bell-shaped or convex, yellowish-brown to clay-brown, densely radially fibrillose, with enrolled woolly fringed margin.
Stem: 6-7 cm high, 5-8 mm wide, same colour as, but paler than cap, fibrillose or scaly, with prominent ring-zone of whitish cottony fibrils often becoming black due to trapped spores.
Gills: adnexed or adnate, almost black, mottled, with white edge bearing watery droplets in damp weather.
Spore-print: almost black.
Habitat: tufted from roots or buried wood, in deciduous woodland. Spring to autumn. Common

HYPHOLOMA

Twelve species, both terrestrial and lignicolous, growing singly, gregariously or tufted.

HYPHOLOMA FASCICULARE
Sulphur Tuft
Syn: *Naematoloma fasciculare*
Cap: 2-4 cm diam., shallowly bell-shaped to shallowly convex, sulphur-yellow often with tawny flush at centre, margin with dark fibrillose remnants of veil.
Stem: 4-7 cm high, 4-8 mm wide, same colour as cap, sometimes flushed brown below, with poorly developed purplish-brown fibrillose ring-zone near apex.
Gills: sinuate, sulphur-yellow becoming olive.
Flesh: yellow with bitter taste.
Spore-print: purplish-brown.
Habitat: tufted at base of deciduous and coniferous tree-stumps. Autumnal, but sporadically throughout the year. Very common.

HYPHOLOMA SUBLATERITIUM
Syn: *Naematoloma sublateritium*
Cap: 4-7 cm diam., convex to flat, brick-red, paler and yellowish toward margin which often bears fibrillose remnants of veil.
Stem: 6-9 cm high, 8-10 mm wide, fibrillose, yellowish above reddish-brown below with fibrillose ring-zone near apex.
Gills: adnate or sinuate, yellowish becoming greyish-violet to purplish-brown.
Flesh: yellowish, but reddish-brown in base of stem, taste mild.
Spore-print: purplish-brown.
Habitat: tufted at base of deciduous tree-stumps. Autumnal. Occasional.
Recognized by brick-red cap and tufted habit at base of trunks or stumps. *H. fasciculare* differs in sulphur-yellow colour, olive gills and bitter taste. In addition the stem and flesh of *H. fasciculare* become bright orange with ammonia whereas in *H. sublateritium* the colour change is to bright yellow.

BOLETES

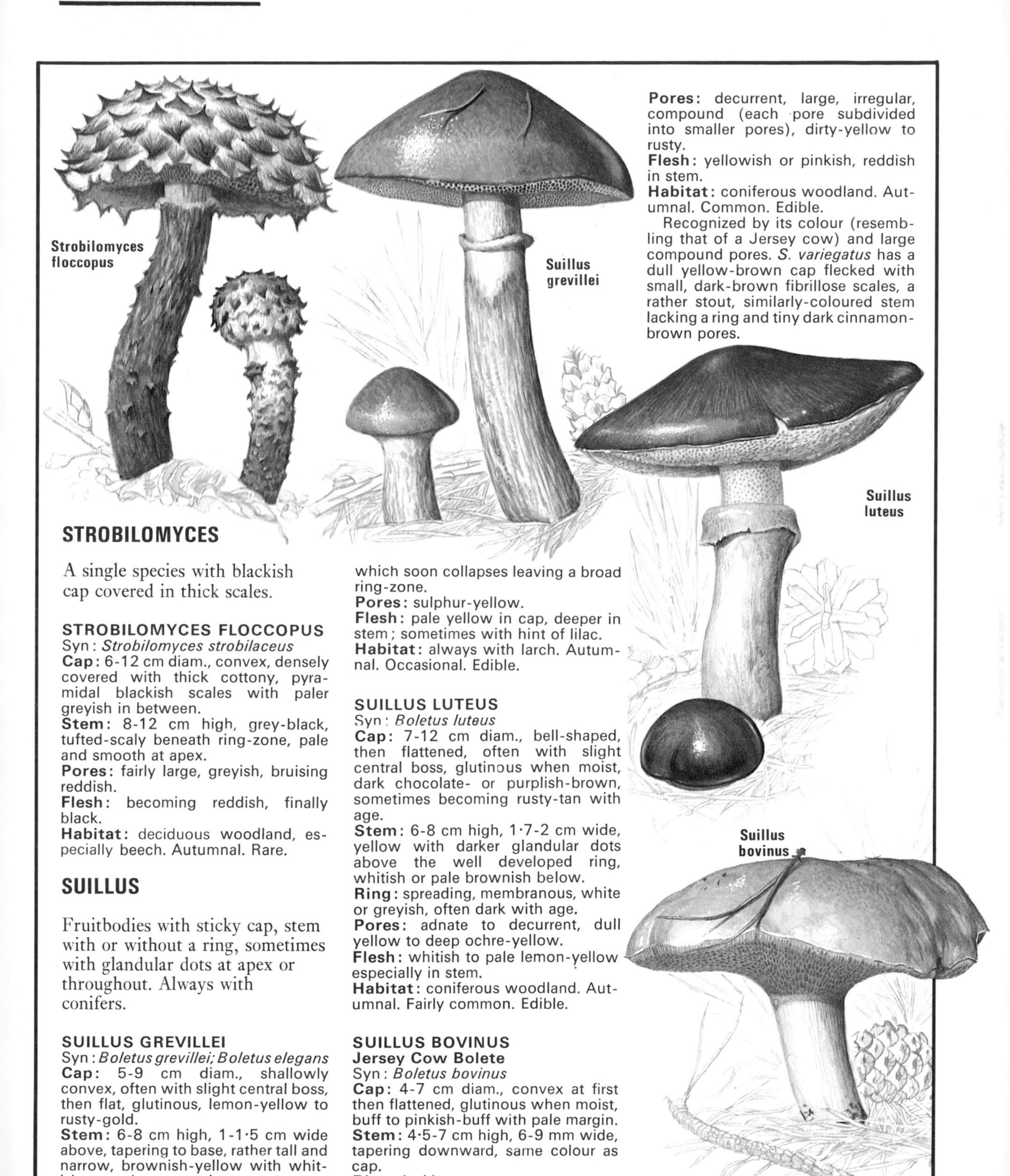

STROBILOMYCES

A single species with blackish cap covered in thick scales.

STROBILOMYCES FLOCCOPUS

Syn: *Strobilomyces strobilaceus*
Cap: 6-12 cm diam., convex, densely covered with thick cottony, pyramidal blackish scales with paler greyish in between.
Stem: 8-12 cm high, grey-black, tufted-scaly beneath ring-zone, pale and smooth at apex.
Pores: fairly large, greyish, bruising reddish.
Flesh: becoming reddish, finally black.
Habitat: deciduous woodland, especially beech. Autumnal. Rare.

SUILLUS

Fruitbodies with sticky cap, stem with or without a ring, sometimes with glandular dots at apex or throughout. Always with conifers.

SUILLUS GREVILLEI

Syn: *Boletus grevillei; Boletus elegans*
Cap: 5-9 cm diam., shallowly convex, often with slight central boss, then flat, glutinous, lemon-yellow to rusty-gold.
Stem: 6-8 cm high, 1-1·5 cm wide above, tapering to base, rather tall and narrow, brownish-yellow with whitish membranous ring near apex which soon collapses leaving a broad ring-zone.
Pores: sulphur-yellow.
Flesh: pale yellow in cap, deeper in stem; sometimes with hint of lilac.
Habitat: always with larch. Autumnal. Occasional. Edible.

SUILLUS LUTEUS

Syn: *Boletus luteus*
Cap: 7-12 cm diam., bell-shaped, then flattened, often with slight central boss, glutinous when moist, dark chocolate- or purplish-brown, sometimes becoming rusty-tan with age.
Stem: 6-8 cm high, 1·7-2 cm wide, yellow with darker glandular dots above the well developed ring, whitish or pale brownish below.
Ring: spreading, membranous, white or greyish, often dark with age.
Pores: adnate to decurrent, dull yellow to deep ochre-yellow.
Flesh: whitish to pale lemon-yellow especially in stem.
Habitat: coniferous woodland. Autumnal. Fairly common. Edible.

SUILLUS BOVINUS
Jersey Cow Bolete

Syn: *Boletus bovinus*
Cap: 4-7 cm diam., convex at first then flattened, glutinous when moist, buff to pinkish-buff with pale margin.
Stem: 4·5-7 cm high, 6-9 mm wide, tapering downward, same colour as cap.
Ring: lacking.
Pores: decurrent, large, irregular, compound (each pore subdivided into smaller pores), dirty-yellow to rusty.
Flesh: yellowish or pinkish, reddish in stem.
Habitat: coniferous woodland. Autumnal. Common. Edible.

Recognized by its colour (resembling that of a Jersey cow) and large compound pores. *S. variegatus* has a dull yellow-brown cap flecked with small, dark-brown fibrillose scales, a rather stout, similarly-coloured stem lacking a ring and tiny dark cinnamon-brown pores.

LECCINUM

Fruitbody boletoid, with dry, convex cap. Flesh often changing colour when cut.

LECCINUM SCABRUM

Syn: *Boletus scaber*
Cap: 5-10 cm diam., convex to shallowly convex, grey-brown with an almost granular-mottled effect under a lens.
Stem: 8-12 cm high, 2-3 cm wide, tall, cylindrical, often enlarged below, white but rough with conspicuous black flocci.
Pores: almost free, minute, dingy buff, bruising yellowish-brown.
Flesh: white, unchanging or faintly pink.
Habitat: with birch, especially on heaths. Autumnal. Very common. Edible.

LECCINUM VERSIPELLE

Syn: *Boletus versipellis; B. testaceo-scaber; L. testaceo-scabrum*
Cap: 6-13 cm diam., convex with fringe-like margin which over-reaches the pores, tawny-orange.
Stem: 9-14 cm high, 2·5-3 cm wide, tall, cylindrical, slightly enlarged below, white, rough-dotted with black flocci.
Pores: almost free, minute, greyish.
Flesh: white, becoming greyish-mauve, but blue-green at base of stem, finally blackening.
Habitat: with birch, especially on heaths. Autumnal. Common. Edible.

The bright tawny-orange cap and black dotted stem make identification easy. *L. aurantiacum* is similar but is associated with poplar

BOLETUS

Fruitbody fleshy. Flesh often bruising indigo blue or blue-green when broken. Cap large to very large, convex, dry or humid but never conspicuously sticky, surface felty or downy, sometimes cracking. Stem thick, cylindrical or conspicuously bulbous, without a ring.

BOLETUS BADIUS

Syn: *Xerocomus badius*
Cap: 6-10 cm diam., shallowly convex, dark bay to chocolate-brown, slightly sticky in wet weather, shiny when dry but softly downy at margin.
Stem: 7-8 cm high, 1·5-2 cm wide, pale brown with darker streaks.
Pores: adnate or adnexed, small, cream to lemon-yellowish, turning blue-green when bruised.
Habitat: deciduous and coniferous woodland. Autumnal. Common. Edible.

Recognized by dark brown cap and pale creamy pores which rapidly become blue-green to touch.

BOLETUS CHRYSENTERON

Syn: *Xerocomus chrysenteron*
Cap: 5-7 cm diam., convex, madder-brown sometimes with olive tint, cracking in a chequered manner showing pale pinkish flesh in between; surface minutely felty.
Stem: 6-8 cm high, 1 cm wide, yellowish streaked red below.
Pores: adnate to adnexed, large angular, at first pale dull yellow then olive-yellow.
Flesh: pink under cuticle, yellowish elsewhere often reddish in stem, turning slightly blue when cut.
Habitat: deciduous woodland. Autumnal. Very common.

Recognized by madder-brown cap cracking to show pink flesh. *B. armeniacus* has a peach-coloured cap, while in *B. rubellus* it is blood-red, and in *B. pruinatus* dark purplish-bay to almost blackish with velvety look and with brighter yellow pores. In none of these species does the cap crack in such a conspicuously chequered manner.

B. subtomentosus, a common species of heathy areas, differs from *B. chrysenteron* in its velvety olive-tan coloured cap bruising dark brown on handling, the lack of pink colour under the cuticle which seldom cracks, a buff stem ornamented with coarse, branching, rusty ribs and large bright golden yellow pores, bluing slightly to touch.

BOLETUS ERYTHROPUS

Cap: 7-11 cm diam., convex, dry, minutely downy, dark bay-brown to red-brown.
Stem: 8-11 cm high, 1·5-2·5 cm wide, yellow densely covered with very minute red granular-dots.
Pores: free, minute, blood-red.
Flesh: yellow, instantly indigo-blue like rest of fruitbody when cut or bruised.
Habitat: deciduous woodland. Autumnal. Fairly common. Edible despite startling colour change of flesh.

Recognized by dark brown cap, blood-red pores, red-dotted stem and indigo blue colour when handled or in cut flesh. *B. luridus* is similar but has conspicuous red-network on stem. *B. queletii* is best distinguished by beetroot colour in the base of the stem.

BOLETUS EDULIS
Penny Bun Bolete

Cap: 10-16 cm diam., strongly convex, usually chestnut brown.
Stem: 6-12 cm high, 2·5-4 cm wide, cylindrical but sometimes conspicuously swollen at base to 10 cm wide; pale with whitish network at least at apex.
Pores: adnexed, whitish becoming greenish-yellow.
Flesh: whitish unchanging, sometimes faintly pink in cap.
Habitat: deciduous and coniferous woodland. Late summer to autumn. Fairly common. Edible and excellent. (The 'Cep' or 'Steinpilz' of continental gourmets and a major constituent of many 'mushroom' soups).

The penny-bun coloured cap, whitish pores and pale stem with whitish net are distinctive. *B. aestivalis* (Syn: *B. reticulatus*) is separated by its pale straw to pale snuff-brown cap. *B. aereus* has blackish cap lacking red tints and *B. pinicola* red-brown cap.

BOLETUS PARASITICUS
Parasitic Bolete

Cap: 3-5 cm diam., convex then flat, minutely felty, tawny with olive tint.
Stem: 3-6 cm high, 6-8 mm wide, slender curved, yellowish streaked with rusty-red.
Pores: adnate, golden often with red blotches.
Flesh: yellow, reddish in stem, unchanging.
Habitat: one or more parasitic on the common Earth Ball (*Scleroderma citrinum*) in heathy or sandy localities. Autumnal. Rare.

Distinctive due to the unique habitat; the only parasitic bolete but in Japan *Xerocomus astereicola* grows on the earth star *Astraeus hygrometricus*.

Most boleti are liable to attack by the mould *Sepedonium chrysospermum*, which envelops them in a whitish mycelium and eventually produces a bright yellow powdery mass of spores, chiefly over the pores of the bolete.

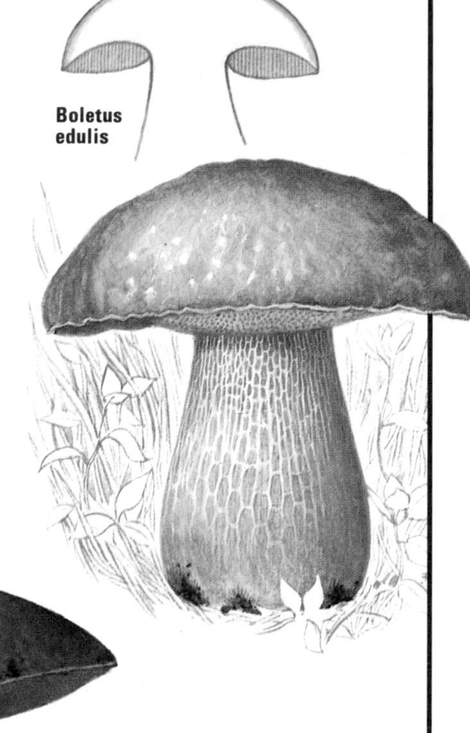

Boletus erythropus

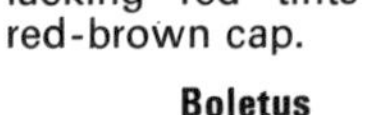

BOLETUS SATANAS
The Devil's Boletus

Cap: 10-18 cm diam., convex with enrolled margin, pale-greyish.
Stem: 7-12 cm high, enormously swollen below up to 12 cm wide, yellow above, red below with conspicuous red network.
Pores: free, minute, blood-red.
Flesh: pale yellow, turns faintly blue in stem apex and over tubes.
Habitat: beechwoods on chalk. Autumnal. Rare. POISONOUS.

The large size, pale grey cap, enormously swollen stem with red net and the red pores are the salient features. In *B. satanoides* and *B. purpureus* the flesh becomes vivid blue when cut and the cap develops reddish or pinkish tones toward the margin. *B. calopus* differs in having yellow pores. Like *B. satanas* it has a red base to the stem but net is white or pinkish. It is the most common of the group of species discussed above.

HOW TO TAKE A SPORE-PRINT

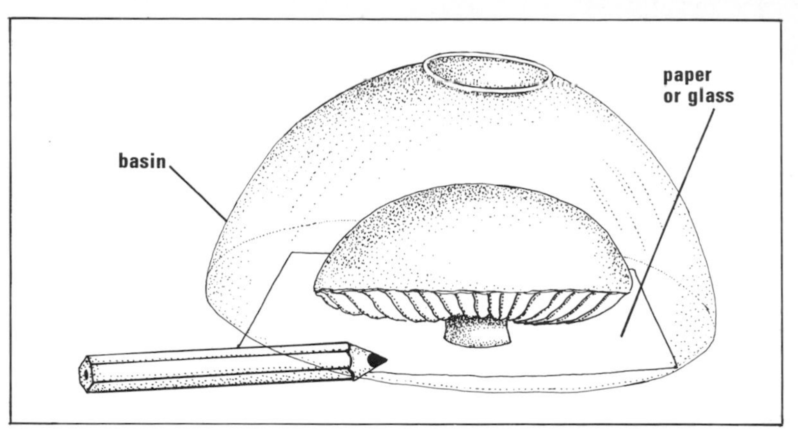

Cut off the stalk as near the cap as possible and place the cap (gills downward) on paper, or preferably glass, and cover with a basin or jar for 12 to 24 hours. If the gills are white, the cap is best put on black paper, but if the gills are coloured, white paper should be used. An exact replica of the gill arrangement, formed by the deposit of spores, will result. It is often necessary to support the basin on a pencil at one side to avoid undue humidity which might interfere with spore deposition. Spore prints of very delicate or tiny species should be taken in tins to prevent drying out, but even here it is often necessary to leave the lid not quite closed. Spore deposits, which may be white (sometimes pale cream or faintly pink), brownish-pink, brown, purplish or black, can be fixed by spraying with hair lacquer.

GLOSSARY

adnate (of gills) broadly attached to stem by their entire width (see page 5).
adnexed (of gills) narrowly attached to stem by less than their entire width (see page 5).
adpressed closely pressed to surface.
agarics gill-bearing fleshy fungi, mushrooms.
apical at the tip.
basidium (pl. basidia) spore-bearing organ of those fungi covered in this book.
campanulate broadly bell-shaped.
cartilaginous tough, firm, not easily snapped.
chlamydospores spores with thick walls that are capable of surviving unfavourable growing conditions.

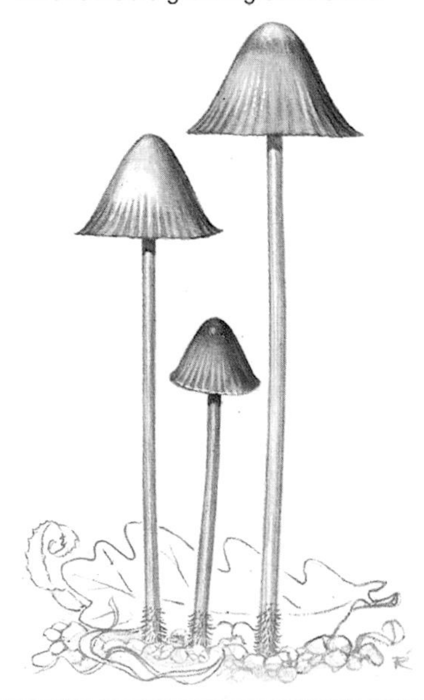

concolorous of the same colour.
cortina a zone of cobweb-like threads near the top of the stem in some agarics.
crispate (of cap margin) wavy.
cristate finely divided, fringed, crested.
cuticle outermost layer of cap or stem.
deciduous (of ring on stem) soon falling away.
decurrent (of gills) running down the stem (see page 5).
dehiscence (of fruitbodies) opening at maturity by a pore or splitting open to facilitate liberation of spores.
dentate (of cap margin) fringed with tiny tooth-like fragments.
dichotomous branching equally into two like a tuning fork.
fasciculate in tufts or bundles.
fastigate having the growth form of a Lombardy Poplar, i.e. with many densely-crowded, erect branches.
fibrillose formed of threads or fibrils.
flocci tiny, woolly scales.
glabrous naked, smooth.
gregarious growing several together or in swarms.
hispid covered in stiff bristles or hairs.
hydrated water-soaked.
hyphae branching threads that form the mycelium and fruitbody.
lignicolous growing on wood.
lilaceous lilac-coloured.
membranous thin, skin-like.
mycelium sterile, felt-like or cobwebby vegetative stage of a fungus.
papilla small, nipple-like outgrowth.
pruinose having a bloom like that of a ripe plum; as if covered with a fine layer of frost.
pustule a tiny cushion-shaped outgrowth, hence pustular.
pyriform pear-shaped, with a head which gradually narrows below into a stalk-like base.
ring or ring-zone the remnants of the partial veil on the stem of certain gill and pore fungi.
rugulose uneven, with fine, rather indistinct wrinkles.
saccate sac-like or cup-shaped.
sessile without a stalk.
sinuate (of gills) notched just before joining stem (see page 5).
squamulose bearing tiny scales.
stellate star-shaped.
striae fine lines that are radial on a cap surface and longitudinal on a stem.
striate marked with striae; (of cap margin) indicating that it is radially lined often due to the gills showing through.
supine (of fruitbodies) prostrate, without a cap, forming paint-like patches on trunks or branches.
tomentose having a felt-like texture.
tuberculate bearing small warts.
umbilicate shaped like a navel.
veil a sheath-like covering especially appertaining to the agaric fruitbody.
velar derived from a veil.
viscid sticky.
volva sac- or cup-like structure at the base of the stem in some agarics.
volval derived from a volva.

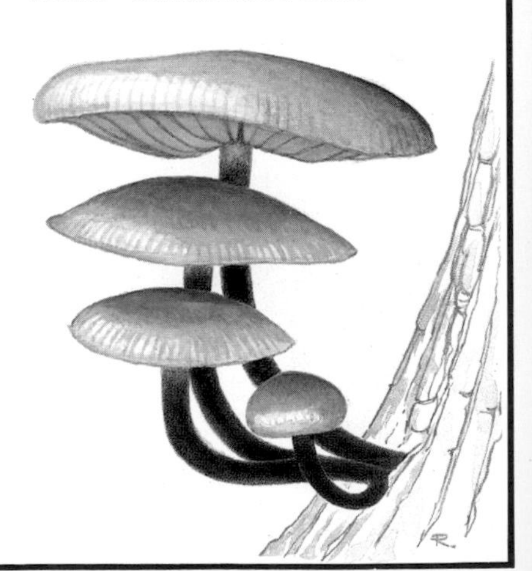

INDEX

Common Names

Amethyst Deceiver 18
Antabuse Ink Cap 30

Blewit 17
Blusher 7
Brittle Gills 14

Chanterelle 10
Clouded Clitocybe 13
Coconut Smelling Milkcap 11

Death Cap 6
Deceiver 18
Devil's Boletus 37

Fairy Cake Fungus 29
Fairy-Ring Champignon 19
False Chanterelle 10
False Death Cap 6
Field Mushroom 4, 5, 31
Fly Agaric 6
Foxy Spot 18

Gas Tar Fungus 20

Honey Fungus 9
Horn of Plenty 10
Horse Mushroom 5, 31

Ink Caps 30

Jersey Cow Bolete 35

Lawyer's Wig 30
Leathery Mycena 22
Liberty Cap 34
Little Japanese Umbrella 30
Long Root 9

Magpie 30
Milk Caps 11
Milking Mycena 23
Miller 25

Oyster Fungus 24

Parasitic Bolete 37
Parasol Mushroom 8
Penny Bun Bolete 37
Pick-a-Back Fungi 24
Plums and Custard 17
Poached Egg Fungus 9

Roll Rim 29
Rooting Shank 9

St George's Mushroom 20
Sea Green Clitocybe 13
Shaggy Ink Cap 30
Shaggy Parasol 8
Shaggy Pholiota 26
Sickener 15
Spindle Shank 41
Sulphur Tuft 34

Tawny Grisette 7
Trooping Crumble Cap 31
Trumpet of Death 10

Ugly One 12

Velvet Shank 17
Verdigris Fungus 32

Weeping Widow 34
Wood Blewit 17
Wood Woolly Foot 19

Yellow Staining Mushroom 32

Botanical Names

Agaricus 31, 32
arvensis 4, 5, 31
bisporus 4
bitorquis 31
campestris 4, 5, 31
edulis 31
haemorrhoidarius 32
langei 32
meleagris 32
placomyces 32
silvaticus 32
xanthodermus 32
Agrocybe 28
praecox 28
Amanita 4, 6
citrina 6
crocea 7
excelsa 7
fulva 7
mappa 6
muscaria 6
pantherina 7
phalloides 6
rubescens 7
spissa 7
Amanitopsis fulva 7
Anellaria
semiovata 33
separata 33
Armillaria 9
mellea 9
mucida 9
Asterophora
lycoperdoides 14, 24
parasitica 14, 24
Astraeus hygrometricus 37

Bolbitilus vitellinus 28
Boletus 36
aereus 37
aestivalis 37
armeniacus 36
badius 36
bovinus 35
calopus 37
chrysenteron 36
edulis 37
elegans 35
erythropus 37
grevillei 35
luridus 37
luteus 35
parasiticus 37
pinicola 37
pruinatus 36
purpureus 37
queletii 37
reticulatus 37
rubellus 36
satanas 37
satanoides 37
scaber 36
subtomentosus 36
testaceo-scaber 36
versipellis 36

Cantharellus 10
aurantiacus 10
cibarius 10
Clitocybe 13
aurantiaca 10
discolor 14
flaccida 14
gilva 14
infundibuliformis 14
inversa 14
nebularis 13, 25
odora 13
phyllophila 14
splendens 14
vibecina 14
Clitopilus 25
prunulus 25
Collybia 18
confluens 19
dryophila 19
fusipes 18
maculata 18
mucida 9
peronata 19
Coprinus 30
acuminatus 30
atramentarius 30
comatus 30
disseminatus 31
picaceus 30
plicatilis 30
Cortinarius 17, 27
armillatus 27
pseudosalor 27
sanguineus 27
semisanguineus 27
Craterellus 10
cornucopioides 10
Crepidotus 29
calolepsis 29
mollis 29
Cystoderma 9
amianthina 9

Entoloma 25
clypeatum 25

Flammula penetrans 26
Flammulina velutipes 17

Galerina 27
mutabilis 27
Gomphidius 33
rutilus 33
viscidus 33
Gymnopilus 26
junonius 26
penetrans 26

Hebeloma 29
crustuliniforme 29
Heterobasidion
annosum 11
Hygrophoropsis 10
aurantiaca 10
Hygrophorus 20
coccineus 20
conicoides 21
conicus 21
miniatus 20
nigrescens 21
niveus 21
pratensis 21
puniceus 20
russo-coriaceus 21
splendidissimus 21
virgineus 21
Hypholoma 34
candolleanum 33
fasiculare 34
sublateritium 34
velutinum 34

Inocybe 28
asterospora 28
geophylla 28
geophylla lilacina 28

Laccaria 18
amethystea 18
amethystina 18
laccata 18
proxima 18
Lacrymaria 34
velutina 34
Lactarius 11
britannicus 13
chrysorrheus 13
controversus 11
deliciosus 13
glaucescens 11
glyciosmus 11, 12
hepaticus 12, 13
mairei 11
piperatus 11
plumbeus 12
pubescens 11
quietus 12
rufus 12
tabidus 13
torminosus 11
turpis 12
uvidus 11
vellereus 11
vietus 12
Leccinum 36
aurantiacum 36
testaceo-scabrum 36
versipelle 36
Lepiota 8
amianthina 9
cristata 8
procera 8
rhacodes 8
Lepista 17
nuda 17
personata 17
personatum 17
saeva 17
Leptonia 27
serrulata 27

Index

Macrolepiota 8
Marasmius 19
  confluens 19
  dryophilus 19
  peronatus 19
  oreades 19
Mycena 22
  crocata 23
  epipterygia 22
  epipterygioides 22
  galericulata 22
  galopus 23
  haematopus 23
  inclinata 22
  leucogala 23
  polygramma 21
  rorida 22
  sanguinolenta 23
  viscosa 22

Naematoloma
  fasciculare 34
  sublateritium 34
Nyctalis 24
  asterophora 24
  parasitica 24

Omphalina 13
Oudemansiella 9
  badia 9
  longipes 9
  mucida 9
  radicata 9

Panaeolus 33
  semiovatus 33
  separatus 33
Paxillus 29
  involutus 29
Pholiota 26
  aurivella 26
  mutabilis 27
  praecox 28
  spectabilis 26
  squarrosa 26
Pleurotellus 29
Pleurotus 23, 29
  cornucopiae 24
  corticatus 23
  dryinus 23
  ostreatus 24
  ostreatus columbrina 24
  sapidus 24
Pluteus 25
  cervinus 25
Psalliota arvensis 31
  bitorquis 31
  campestris 31
  placomyces 32
  xanthoderma 32
Psathyrella 33
  candolleana 33
  disseminata 31
  gracilis 33
  hydrophila 33
Psilocybe 34
  semilanceata 34

Rhodophyllus
  clypeatus 25
Rhodotus palmatus 23
Russula 14
  acrifolia 14
  adusta 14
  aeruginea 16
  atropurpurea 15
  caerulea 15, 16
  claroflava 16
  cyanoxantha 16
  delica 11
  drimeia 15
  emetica 15
  fellea 16
  foetens 16
  fragilis 15
  heterophylla 16
  ionochlora 16
  laurocerasi 16
  mairei 15
  nigricans 14, 24
  ochroleuca 16
  parazurea 16
  queletii 15
  sanguinea 15
  sardonia 15, 16
  sororia 16
  xerampelina 15

Strobilomyces floccopus 35
  strobilaceus 35
Stropharia 32
  aeruginosa 32
  merdaria 32
  semiglobata 32
Suillus 35
  bovinus 35
  grevillei 35
  luteus 35
  variegatus 35

Tephrocybe 18
Tricholoma 20
  album 20
  argyraceum 20
  cingulatum 20
  gambosum 20
  lascivum 20
  nudum 17
  personatum 17
  rutilans 17
  saevum 17
  sulphureum 20
  terreum 20
Tricholomopsis 17
  platyphylla 17
  rutilans 17
Tubaria 28
  furfuracea 28

Volvariella 24
  bombycina 24
  gliocophala 25
  speciosa 25
  surrecta 25

Xerocomus astereicola 37
  badius 36
  chysenteron 36